MICROBIAL BIOPESTICIDES IN INDIA

NIPA GENX ELECTRONIC RESOURCES & SOLUTIONS P. LTD.
New Delhi-110 034

About the Authors

Dr. R. K. Murali Baskaran is working as Principal Scientist (Agricultural Entomology) at ICAR-National Institute of Biotic Stress Management, Raipur, Chhattisgarh and the Principal author of the book. He is specialized in Biological Control of crops. He has completed 32 years of service in Teaching, Research and Extension of Agricultural Entomology. Currently he is working on the development of plant volatile repository for crop pest management. He has two decade experience in teaching and guided 14 students' dissertations. He is in recipient of Young Scientist Fellowship from Tamil Nadu State Council for Science and Technology, Chennai and Best Researcher Award twice from Tamil Nadu Agricultural University, Coimbatore. To his account, he has published around 50 research papers in Biological control. He is acting as reviewer in many international entomology journals.

Dr. J. Sridhar is working as Senior Scientist (Agricultural Entomology) at ICAR-National Institute of Biotic Stress Management, Raipur, Chhattisgarh. His specialization is on insect vector interactions. Development of indigenous protocols for detection of potato viruses, aphids map of potato, transmission efficiencies of five vector aphids with respect to PVY and PLRV and whole genome sequencing of *Aulacorthum solani* are his noteworthy achievements when he worked at ICAR-Central Potato Research Institute, Shimla. He has done countrywide mapping of genetic groups of *Bemisia tabaci* and thrips populations in Chhattisgarh. Currently, he is working on emerging and re-emerging pests in Conservation Agriculture production systems. He is involved in teaching of mastoral students in affiliation with ICAR-IARI, New Delhi and received a medal from Indian Potato Association for best research paper. To his credit, he has published more than 27 research papers in peer reviewed journals and acting as reviewer of four international journals.

Dr. Mallikarjuna Jeer is working as Senior Scientist (Agricultural Entomology) at ICAR-National Institute of Biotic Stress Management, Raipur. He is specialized in host plant resistance and currently is working in Silicon mediated resistance against insect pests, identifying novel genetic resources and genes for imparting resistance in okra, brinjal and *Vigna* sp., and AICRP Nematodes as Principal Investigator. He has published many of his research outputs in the form of research papers in peer reviewed journals, book chapters, technical bulletins, extension folders etc. He is a recipient of many recognitions and awards including young scientist award and Fellow of Entomological Society of India.

Dr. Kaushik Banerjee is a Principal Scientist from ICAR-National Research Centre for Grapes, Pune. He is heading the National Reference Laboratory on pesticide residues and mycotoxins in India. His area of research focuses on the development of efficient analysis methods for the sensitive and confirmatory estimation of pesticide residues and mycotoxins in agricultural and food matrices and risk assessment studies for the fixation of crop-specific maximum residue limits. Dr. Banerjee's extensive contributions to science and the community have earned him numerous national and international laurels. He received the prestigious Harvey W. Wiley Award of AOAC INTERNATIONAL in 2017 and the Recognition Award of the National Academy of Agricultural Sciences (NAAS) in 2019. Earlier, he was named as a Fellow by the Royal Society of Chemistry (FRSC), National Fellow-Indian Council of Agricultural Research, and Fellow-NAAS.

Dr. P. K. Ghosh PhD topped in M.Sc. and Ph.D., with First Class First (90%) & Gold Medal in GBPUA&T, Pantnagar, He started his career as scientist in Indian Council of Agricultural Research (ICAR) in April 1993, completed 28 years' service (11 years as Research Management position and 14 years Professor/Principal Scientist) and is presently working as Founder Director and Vice Chancellor ICAR-National Institute of Biotic Stress Management (NIBSM)-Deemed to be University, Raipur, Chhattisgarh, India. Earlier he demonstrated leadership as National Coordinator, National Agricultural Higher Education Project-World Bank aided project, ICAR, New Delhi (2017-20), Director, Indian Grassland and Fodder Research Institute, Jhansi (2012-17) and Head, Crop Production Division Indian Institute of Pulses Research (2009-2012). His pioneering works on climate change, carbon sequestration, Resource use efficiency, conservation agriculture crop diversification, soil water conservation, Integrated Farming System (IFS) and coordination of higher education in 14 agricultural universities in India to bring excellence in quality of PG education program have been considered as outstanding contributions benefitting large numbers of farming community, scientists and students across the globe, as result he has been identified in November 2020 as one of the top two percent of agricultural scientists at global level based on the database analysed by Stanford

University, USA. With his initiative PG program on biotic stress management at NIBSM has begun in 2020 affiliated with IARI, New Delhi. He has implemented National Initiative on Fodder Technology Demonstration throughout the country through 100 KVKs, established Adarh Charagram and grassland in 6 states in 657-acre area and promoted 17 Gousala in UP and Rajasthan for ensuring fodder security and livestock productivity. He is recipient of 19 national awards including M.S. Randhawa Memorial Award for best Administration, Excellence in Science Award and Sardar Patel Outstanding–Best ICAR Institute Award for leadership and also Fellow of National Academy of Sciences, India and National Academy of Agricultural Sciences (NAAS).One of his international book entitled "Carbon Management in Tropical and Sub-Tropical Terrestrial System" (published by Springer) is being referred globally and nationally to address Carbon Neutrality, which is one of the important points of SDG 2030. His another book "Grassland: A Global Resource Perspective" has become very popular to the International livestock and grassland working groups. He showed scientific leadership as Chairman by organizing 23rd International Grassland Congress (IGC) held first time in India with 450 international delegates and still acting as executive member of IGC continuing committee representing South-East Asia. He served as sectional president, Agricultural section under the general presidentship of Hon'ble Prime Minister of India in the centenary year (2013) of Indian Science Congress. Dr Ghosh holds many important positions nationally and internationally, few of them are Member, DST task force on Climate change on forestry, Chairman, National fodder planning committee, Editor-in-Chief, National Academy of Science, India, Facilitator, National Conference on Doubling Farmer Income (DFI) and Chairmen/Coordinator/Convenor of many national seminar/workshop/ conference/brainstorming, Member, Gujarat Institute of Desert Ecology, Member, Task Force on DUS guideline, Member, Board of Management /Academic Council/PG faculty of 9 Universities including Country expert/coordinator of many international projects funded by CGIAR/FAO/World Bank. Considering his academic, administrative and research excellence he was selected as co-opted member of DFI Committee, constituted by PMO and contributed to bring 14 volumes on DFI which are available in website of Ministry of Agriculture and Farmers Welfare and many actionable points and recommendation are being executed in different states. He also formulated four policy papers for the country and delivered 33 lead /Keynote Lectures in National and International conferences. He published 20 books and 161 research papers in the high impact factor journals with the total citation of 7103, h-index of 39 and i-10 index of 83 and total publication of 283.

MICROBIAL BIOPESTICIDES IN INDIA

R. K. Murali Baskaran
Principal Scientist
School of Crop Health Management Research
ICAR-National Institute of Biotic Stress Management
Raipur, Chhattisgarh

Sridhar, J.
Senior Scientist
School of Crop Health Biology Research
ICAR-National Institute of Biotic Stress Management
Raipur, Chhattisgarh

Mallikarjuna, J.
Senior Scientist
School of Crop Resistance System Research
ICAR-National Institute of Biotic Stress Management
Raipur, Chhattisgarh

Kaushik Banerjee
Principal Scientist
ICAR-National Research Centre for Grapes
Pune, Maharashtra

P. K. Ghosh
Director and Vice Chancellor
ICAR-National Institute of Biotic Stress Management Baronda
Raipur, Chhattisgarh

NIPA GENX ELECTRONIC RESOURCES & SOLUTIONS P. LTD.
New Delhi-110 034

NIPA GENX ELECTRONIC RESOURCES & SOLUTIONS P. LTD.

101,103, Vikas Surya Plaza, CU Block
L.S.C. Market, Pitam Pura, New Delhi-110 034
Ph. +91 11 27341616, 27341717, 27341718
E-mail: newindiapublishingagency@gmail.com
www: www.nipabooks.com

For customer assistance, please contact
Phone: + 91-11-27 34 17 17
Fax: + 91-11-27 34 16 16
E-Mail: feedbacks@nipabooks.com

ISBN: 978-93-95319-88-1

Composed and Designed by NIPA.

भारत सरकार
कृषि अनुसंधान परिषद विभाग एवं
कृषि एवं किसान कल्याण मंत्रालय, कृषि भवन, नई दिल्ली 11001

GOVERNMENT OF INDIA
DEPARTMENT OF AGRICULTURAL RESEARCH & EDUCATION (DARE)
AND
INDIAN COUNCIL OF AGRICULTURAL RESEARCH (ICAR)
MINISTRY OF AGRICULTURE AND FARMERS WELFARE
KRISHI BHAVAN, NEW DELH-1110 001
Tel.: 23382629; 23386711 Fax: 91-11-23384773
E-Mail: dg.icar@nic.in

डॉ. हिमांशु पाठक
सचिव, एवं महानिदेशक
DR HIMANSHU PATHAK
SECRETARY (DARE)
& DIRECTOR GENERAL (ICAR)

Foreword

In India, the crop losses due to pests and diseases are estimated to the tune of 30-40% under field condition and 9-10% post-harvest. The estimates suggest that pathways for introduction of invasive biotic stress have increased post globalization and free trade policy leading to losses associated with invasive insects, plants, and pathogens worth up to $ 1.4 trillion annually. These biotic stresses are compounded by climate change influences. During Green Revolution and thereafter, the frequent and high application of chemicals and pesticides in agriculture, approximately 500 insect and related arthropod species are reportedly developed resistance against major groups of chemical pesticides, besides pesticide and chemicals loads in agriculture produces and outbreaks and resurgence of secondary pest etc.

The policy reorientation with strict regulations and ban on some of the hazardous chemical pesticides in the recent past and consumers awareness and preference for healthy agricultural products have put responsible focus on 'Greener Technologies' under environment friendly strategic framework. The 'National Policy for Farmers' (NPF), 2007 emphasised for increasing the organic agriculture which pushed the promotion of microbial biopesticides in India. Although the present policy ecosystem for greener technologies are supportive to the biopesticides, the factors such as high initial cost in identification and development of organisms, lack of subsidies to the small scale manufacturers, complex registration protocols, interest of multinational companies in new pesticide and associated business etc., are some hurdles in rapid growth of biopesticide industries and large scale utilization.

It is heartening to learn that currently 970 microbial biopesticide formulations from 15 microbial species are registered in India and 31 new microbial biopesticides developed by various ICAR institute against crop pests and pathogens are at various stages of registration and commercialization. I congratulate the authors for bringing out a compilation on 'Microbial Biopesticides in India' which shall be a useful reference book for policy makers, researchers, students and all other holders.

Dated: 6th September, 2022
Place: New Delhi

(Himanshu Pathak)

Preface

In Indian agriculture, the widespread application of chemical pesticides to protect crops from biotic stresses has become a regular practice, with several unintended negative impacts on crops, people, animals, soil, water bodies, non-target creatures, and their surrounding environments. By 2050, the world's population is expected to increase to 9 billion people, necessitating a review and revision of current programmes for a 70% increase in food production. The government's strict regulations on chemical pesticides have created a favourable environment and provided a path for alternate, environmentally acceptable methods of managing biotic stress. Microbial biopesticide has been identified as an emerging tactic that is rapidly expanding in the context of plant protection in India. The appropriate modifications and simplifications to registration standards of biopesticides made by the Government, increase of the amount of land used for organic farming, subsidies to investors in the biopesticide industry, and other changes that attracted the attention of manufacturers. Additionally, stakeholders are concentrating on the advantages of biopesticides by raising knowledge of the value of high-quality goods that promote a healthy way of life and likelihood. Research and development efforts to improve the kill rate, shelf life, etc., of biopesticides are an added benefit for the steady growth of industry both internationally and in India. A steady rise in demand and consumption of biopesticides in India is a good sign for their ability to compete with or even surpass the market for chemical pesticides between 2040 and 2050.

A book titled "Microbial Biopesticides in India" consisting of 11 chapters with a focus on the need for biopesticide in plant protection, formulations, nano-biopesticide, genetic engineering, demand, consumption, and market, as well as government initiatives, awareness by growers, driving strategies and set-back to enhance the biopesticide market in India was written by referring latest literature, dailies, review papers etc., in order to bring the Indian perspectives on microbial biopesticide to the most common platform for the benefit of readers, learned faculties and colleagues, corporate, stakeholders, students, youngsters etc.

We, the writers, have been involved in the compilation of peer-reviewed data on the growth and development of biopesticide in India over the past two decades. Comparisons between India's policies and those of other industrialized nations' regulatory frameworks for biopesticide science include a number of concrete steps that the Indian Government has previously taken or plans to take. The coordinated work output by all authors in carefully gathering and compiling a variety of information from numerous sources is appreciated.

Editors

Contents

Introduction

The "Green Revolution" (GR) was brought about by the use of numerous inorganic outputs, such as pesticides, fertilisers, high input responsive cultivars, etc., in Indian agriculture, which greatly increased crop production and productivity. The stakeholders were motivated by the proportionate rise in yield indices to use numerous inorganic inputs carelessly, which had negative effects on soil quality, crop output quality, environmental pollution, human and animal health, etc. Inappropriately using synthetic chemical pesticides on a large scale to protect crops has led to a number of threats, such as insecticide residues in crop products, outbreaks, resurgence, the creation of secondary pests, insecticide resistance, and more. More than 500 insect and related arthropod species have been found to become resistant to significant classes of chemical insecticides.

The use of modern agricultural inputs needs to be reconsidered because the world's population is projected to increase to 9 billion people by 2050, which will result in an additional demand for food of about 70%. Despite this, global agriculture is still in the process of recovering from a number of negative effects. Wherever human action is used to correct the situation and return it to normal, climate change has been reported to amplify the negative effects. The current government's strict controls and rules on the registration, production, marketing, and subsequent field usage of chemical pesticides serve to lower demand and consumption. By raising knowledge among farmers, customers, the general public, etc. about the usage of high-quality goods free of pesticide residues, adulterations, etc., these groups' views were altered, and they were more interested in alternative tactics like "Green Technologies" for plant protection.

Many Government initiatives and programmes which include Sikkim Organic Mission (SOM), National Programme for Organic Production (NPOP), Organic Farming Policy (OFP), Strengthening and Modernizing Pest Management Approach in India (SMPMA), Capital Investment Subsidy Programme (CISP), National Action Plan on Climate Change (NAPCC), National Mission for Sustainable Agriculture (NMSA), "*Paramparagat Krishi Vikas Yojana*" (PKVY), Soil Health Management (SHM), Zero Budget Natural Farming (ZBNF) etc., are supportive for the scope of using 'Microbial Biopesticide' in India. Many small-scale industries were drawn to the biopesticide industry by the continued easing of regulation policies, but the initial investment costs for microbe identification, characterization, bioefficacy tests, toxicology tests, registration, commercialization, and other related costs are prohibitive. Based on their prior experiences in international trade, multinational corporations (MNCs) are interested in innovative pesticide chemistries. In order to increase the demand and market for biopesticides, it is recommended that the government should provide adequate subsidies for early investment expenditures, set a fair pricing for biopesticides, and work with multinational corporations to support small-scale companies.

The number of bio-production units has currently increased to 361, of which 141 are in the private sector without GOI grant aids and 38 with GOI grant aids. Moreover, the Ministry of Agriculture and Farmers Welfare has assisted about 35 IPM centers to produce biopesticides since 2010. A total of 98 State Biocontrol Laboratories were established by the State Departments of Agriculture and Horticulture of Gujarat, Uttar Pradesh, Karnataka, Tamil Nadu, Andhra Pradesh and Kerala as well as the production of microbial pesticides by the Institutions of the Indian Council of Agricultural Research.

A total of 970 biopesticides registered in India by Central Insecticide Board and Registration Committee (CIB&RC) under the 1968 Insecticide Act which include microbial biopesticides of *Bacillus thuringiensis* var. kurstaki (42), var. israelensis (22), var. sphaericus (05), var. galleriae (01), *Pseudomonas fluorescence* (196), *Bacillus subtilis* (04), *Trichoderma viride* (289), *T. harzianum* (51), *Ampyliomyces quisqualis* (02), *Beauveria bassiana* (106), *Metarhizium anisopliae* (30), *Verticillium lecani* (93), *Verticillium chlamydosporium* (03), *Helicoverpa armigera* NPV (30) and *Spodoptera litura* NPV (03) and only 38 biopesticidal formulations. Fungal based- (Trichderma sp.) and Pseudomonas based- biopesticides are popular in India consumption-wise while *Bacillus thuringiensis* based formulations are widely used for plant protection of abroad agriculture.

In India, public sectors contribute 70% of the biopesticides production. Major companies are Biotech International Ltd., New Delhi, International Panaacea Ltd, New Delhi, Ajay Biotech (India) Ltd, Pune, Bharat Biocon Pvt. Ltd., Chhattisgarh, Microplex Biotech and Agrochem Pvt., Mumbai, Excel Crop Care Ltd., Mumbai, Govinda Agro Tech Ltd., Nagpur, Jai Biotech Industries, Satpur, Nasik, Ganesh Biocontrol System, Rajkot, Gujarat Chemicals and Fertilizers Trading Company, Baroda, Gujarat Eco Microbial Technologies Pvt. Ltd., Vadodara, Chaitra Agri-Organics, Mysore, Deep Farm Inputs (P) Ltd., Thiruvanandapuram, Kerala, Kan Biosys Pvt. Ltd., Pune, Indore Biotech Inputs and Research Pvt. Ltd., Indore, Romvijay Biotech Pvt. Ltd., Pondichery, Devi Biotech (P) Ltd., Madurai, Tamil Nadu, T. Stanes and Company Ltd., Coimbatore, Tamil Nadu, Harit Bio Control Lab., Yavatmal and Hindustan Bioenergy Ltd., Lucknow. Few Indian companies which work in biopesticde production in collaboration with foreign companies are Lupin Agro-chemicals, Mumbai, Sugar and distillery companies such as KCP Sugar and Industries Corporation Ltd., Andhra Pradesh, Rajshree Sugars and Chemicals Ltd., Tamil Nadu, New Swadeshi Sugar Mills, Bihar, and Bannari Amman Sugars Ltd., Tamil Nadu.

In India, the usage of biopesticides is growing at a faster pace than that of the chemical pesticides. According to the Directorate of Plant Protection, Quarantine and Storage, Ministry of Agriculture and Farmer Welfare, in the last 10 years, consumption of bio-pesticides increased by 23%, while that of chemical pesticides grew only by 2%. At 2020, the compound annual growth rate (CAGR) of global biopesticide market was approximately 3-5% of the total crop protection market while the market was anticipated to grow by 8.64 % at 2023; 9.7% at 2015-2023; 10.3% at 2014-2022; 15% at 2019-2024; 16% at 2020-2025.

Even though the biopesticide sector in India is growing rapidly, issues including slow kill times and short shelf lives are slowing it down. Around the world, researchers are creating recombinant organisms that contain spider, scorpion, and other venoms, adding and deleting genes of interest, and developing nano-biopesticides that have an efficacy that is comparable to chemical insecticides. For the control of important crop pests, a total of 31 fungal and bacterial based biopesticide formulations are under development and at various phases of commercialization. Between 2040 and 2050, the market for biopesticides is anticipated to surpass that for chemical pesticides, becoming one of the key elements of IPM all over the world.

1

Why We Need Biopesticides Some Case Studies of Chemical Pesticides

Abstract

Pesticides are used in most countries around the world to protect agricultural and horticultural crops against damage by pests and diseases. Injudicious use and unintentional poisoning of synthetic pesticides resulted deadly consequences. Exposure to chemical pesticides can have effects that are acute, chronic and long-term. Unregulated misuse of chemical pesticides lead to mobilization of toxic residues across the food chain, increasing bioaccumulation and environmental persistence. Non-target organisms, beneficial insects, land and aquatic animals are badly affected with the excessive use of chemical pesticides. Additionally, chemical pesticide poisoning poses a global concern due to unnatural death caused by mishandling of chemical pesticides. Biopesticide is one of the promising alternatives which can manage menace caused by pests in agriculture, persistency of pesticides, environmental pollutions, toxic and ill effects on non-target species. The development of biopesticides stimulates modernization of agriculture and will, without a doubt, gradually replace chemical pesticides to a great extent.

Keywords: Biopesticides, Residue, Persistency, Environmental pollutions, Glyphosate

Introduction

The world population is expected to reach 9 billion by 2050. This population growth of 2 to 3 billion people over the next 30 years, combined with the changing diets, would result in a predicted increase in food demand of around 70% by 2050 (UNDESA 2009). To feed the burgeoning population, more food and livelihood opportunities from less per capita arable land and water

---

Aditi Kundu, Supradip Saha, Anirban Dutta, and Abhishek Mandal
Division of Agricultural Chemicals, ICAR-IARI, New Delhi

required. Damage caused by insect and pest is one of the major limiting factors for agricultural food grain production. A major portion of expenditure on pesticides is for protecting the crop in the field (Kumar 2013).

Since the discovery of DDT, numerous pesticides (organochlorines, organophosphates, carbamates, pyrethroids, neonicotinoids, etc.) have been developed and used extensively worldwide with few guidelines or restrictions. Indeed, they help control agricultural pests (including diseases and weeds), plant disease vectors, human and livestock disease vectors and nuisance organisms, and organisms that harm other human activities and structures (gardens, recreational areas, etc.). However, many pesticides have been found to be harmful to the environment and human health. Some of them can persist in soils and aquatic sediments, bio-concentrate in the tissues of invertebrates and vertebrates, move up trophic chains, and affect top predators. They have caused adverse effects on soil health, water quality, produce quality and developed problems like insect resistance, genetic variation in plants, toxic residues food and feed. Moreover dependence on chemical pesticides and their indiscriminate use caused several detrimental effects on ecosystems.

Additionally, poisoning by agricultural pesticides is currently an important cause of human morbidity and mortality worldwide, increasing number of farm workers annually exposed to pesticides in developing countries (Jeyaratnam 1990). Developing countries use only 20% of the world's agrochemicals, yet they suffer 99% of deaths from pesticide poisoning (Jeyaratnam and Chia 1994). It has estimated that some form of poison directly or indirectly is responsible for more than one million illnesses worldwide annually. Acute pesticide toxicity is extremely common in developing countries of the Asia-Pacific region, particularly in settings of low education and poor regulatory frameworks. For, deliberate self-poisoning, a plausible range 233,997 to 325,907 with the estimated number of 258,234 probable deaths occurfrom pesticide self-poisoning worldwide each year (Gunnell et al. 2007). Fatality rates of 20% are common and the World Health Organization has estimated that more than 200,000 people die each from pesticide poisoning only (Singh and Unnikrishnan 2006).

Owing to huge adverse environmental impacts of synthetic chemicals, leading to resistance and resurgence of pests, forced to search alternate option for pest management. Further, the increasing public concerns and growing awareness about the potential adverse environmental effects as well as health hazards associated with the use of synthetic plant protection chemicals has prompted search for the technologies and products which are safer for the end users and the environment. Concerns of resistance development in pests and withdrawal of some of the products for either regulatory or commercial reasons, triggered to exploit naturally occurring pesticides.

Biopesticides are environment friendly and safer than classical chemical pesticides. Hence, in the recent years, considerable attention has been paid towards exploitation of biopesticides in protection of food crops/commodities from pest infestations and the associated losses. They are more inclined to use eco-benign natural or herbal products in anticipation of any undesired side effects.Natural occurring phytochemicals have been an excellent option to replace toxic chemical pesticides. It has been speculated that botanical pesticides could reduce the pest resistance problem, thereby often subdue deleterious effects of hazardous chemicals. India has great diversity of flora and fauna. Treasure of bioactive phytochemicals from the diverse plant kingdom need to be exploited to develop newer bioactive molecules. Recent report published by WHO showed more than 21,000 plant species worldwide have tremendous potential for being used in medicinal and phytochemistry. It is estimated that more than 30% of the entire phyto-population possessed active constituents with complex biofunctional characteristics. Bioactive compounds derived from plants have proven to be valuable sources of bioactive secondary metabolites which can seldom be obtained from other sources (Kharshiing 2012).

Case Studies on Acute Toxicity of Chemicals Pesticides

India is an agricultural country with a large rural population (60-80%), where pesticides are easily available and used extensively. Among different pesticides, organophosphates are most commonly used for self-poisoning, but being highly toxic, new compounds with high potency and lower toxicity are being developed continuously. The Poison Information Centre of the National Institute of Occupational Health, in Ahmedabad, reported that organophosphorus (OP) pesticides were responsible for the maximum number of poisonings (73%) among all agricultural chemicals (Dewan and Sayed 1998). Later another study was reported on season-long assessment of acute pesticide poisoning among· cotton growing farmers across three villages in India. The study documented· the serious consequences of pesticide use for the health of farmers, particularly women field helpers who were involved in mixing concentrated chemicals and refilling spraying tanks were as hazardous as direct pesticide application. Of 323 reported events, 83.6% were associated with signs and symptoms of mild to severe poisoning typical of poisoning by organophosphates (Mancini et al. 2005).

Acute intoxications after ingesting glyphosate was reported in the last four decades. Despite low potential toxicity of this herbicide, a number of fatalities and severe outcomes have been reported. Deaths following ingestion of 'Roundup' alone were due to a syndrome that involved hypotension, unresponsive to intravenous fluids or vasopressor drugs, and sometimes pulmonary oedema, in the presence of normal central venous pressure (Talbot et al. 1991). Incidence on acute poisoning in case of suicidal or accidental cases after ingesting glyphosate and its lethal concentration in clinical samples have been published in literature (Hori et al. 2003). Glyphosate-poisoning is characterized by various symptoms such as gastrointestinal symptoms, altered consciousness, hypotension, respiratory distress, metabolic acidosis and renal failure (Tominack et al. 1991; Lee et al. 2000; Roberts et al. 2010). Glyphosate formulation contains surfactants that probably enhance its toxicity. The mortality rate due to glyphosate poisoning is reported at 3.2% in a study included 601 patients with glyphosate acute poisonings. Death was strongly associated with greater age, larger ingestions and high plasma glyphosate concentration >734 μgmL^{-1}. The most common symptoms were oropharyngeal ulceration, nausea and vomiting, mainly due to altered biological parameters of high lactate and acidosis. Respiratory distress, cardiac arrhythmia, hyper-kaleamia, impaired renal function, hepatic toxicity and altered consciousness were the marked observations (Gress et al. 2015; Peillex and Pelletier 2020). Fatalities caused due to cardiovascular shock, cardio-respiratory arrest, haemodynamic disturbance, intravascular disseminated coagulation and multiple organ failure (Zouaoui et al. 2013).

Another case report in Thailand, where poisoning from glyphosate-surfactant herbicide has been displayed with rapid lethal intoxication. For a woman who ingested approximately 500 mL of concentrated Roundup formulation (41% glyphosate as the isopropylamine salt and 15% polyoxyethylene amine) showed glyphosate levels of 3.05 and 59.72 mg/mL in serum and gastric, respectively (Sribanditmongkol et al. 2012). During the re-approval process of glyphosate in Europe, it was mentioned that glyphosate-based products (GBF) were more toxic than glyphosate alone. This phenomenon was attributed to the surfactants and among them, polyethoxylatedtallowamine (POEA) has been suspected to significantly contribute to the toxicity of glyphosate products. In animal data acute oral toxicity of POEA has been suggested to be greater than glyphosate toxicity (Langrand et al. 2020).

Several episodes of mass poisoning by different pesticides have been reported to the Poison Information Centre (PIC) of the National Institute of Occupational Health (NIOH) in Ahmedabad, India, most notably endosulfan, phorate and ethion poisonings. It has been observed that OP poisoning from contaminated food ingestion is all too often treated empirically for food poisoning instead of specific treatment (Patel et al. 2012). A fatal accidental monocrotophos poisoning in adult female by dermal exposure while sleeping has been reported and elevated level of pesticide detected in post-mortem blood and skin by chromatography and spectroscopic techniques (Bodwal et al. 2019).

Endosulfan was one of the highly used organochlorine pesticides, and many poisoning cases have been reported from various regions of the world. In a case study, eighteen incidences of accidental endosulfan poisoning have been reported only from northern India between 1995 and 1997, which occurred after spraying of the pesticide. Analysis of various incriminating factors revealed that accidental overexposure was due to failure to adhere to the instructions for spray either due to ignorance or due to illiteracy (Chugh et al. 1998). Another case study from India revealed poisoning of endosulfan through consumption of endosulfan contaminated water by the entire age group (Srivastava et al. 2009). A survey for 11-year was carried out in various major cities including Ankara in Turkey, insecticides were found to be the most common cause (94%) of fatal pesticide poisoning, with organophosphates such as dichlorvos (25.7%) and organochlorines such as endosulfan (15.7%) being the most common types of pesticides involved (Kır et al. 2013). In Tehran, Iran, another case study showed high level of endosulfan poisoning and the most common culprit was organochlorines (57.1%) insecticide (Akhgari et al. 2018).

Another brief case study reported the inspection from January 2000 to December 2002, revealed 30 positive cases in 2000; 240 positive cases in 2001 and 38 positive cases in 2002. Organophosphorus insecticides were detected as the major component of most samples, representing 63% of the total positive cases and quinalphos is the most abundant pesticide, present in 32 of the 111 positive cases, followed by the herbicide paraquat (Teixeira et al. 2004). Unfortunately, poisoning and fatalities due to endosulfan, a halogenated carbohydrate derivative, have been widely reported in the Indian sub-continent. A fresh 23 cases of endosulfan poisoning have been reported describing symptoms were nausea and vomiting in 17 patients (73.9%), seizures in five patients (21.7%), and dizziness in one patient (4.3%) (Karatas et al. 2006). Two cases of unintentional exposure to endosulfan, one of which presented with neurological manifestations, liver toxicity, and required mechanical ventilation and emergent hemodialysis; the other had only neurological manifestations and liver toxicity, has been reported from a nine-year analysis study in Turkey (Yavuz et al. 2007).

Imidacloprid is a neonicotinoid insecticide developed for commercial use, belonging to the chloronicotinyl nitroguanidine chemical family. Imidacloprid has high potency against insects but with low mammalian toxicity and favorable persistence. On the basis of animal studies, it is classified as "moderately toxic" (class II by WHO and toxicity category II EPAV). Animal studies indicate relatively low toxicity to mammals because they have resistant nicotinic receptor subtypes compared to insects, as well as protection of the central nervous system by the blood brain barrier. Despite wide usage, human exposure experience resulting in toxicity is quite limited. Though imidacloprid regarded as safe for human beings, toxicity can occur through inhalation exposure (Kumar et al. 2013).

Neuropsychiatric symptoms in imidacloprid poisoning have been reported in one case with inhalational exposure mainly due to central nicotinic stimulation (Huang et al. 2006). Cardiovascular manifestations like tachycardia, bradycardia, arrhythmia, and cardiac arrest were also described in different case reports (Wu et al. 2001). There is a paucity of information about human toxicity. Some reports also suggested that the other ingredients in the formulated product could be responsible for causing toxicity. In the same line of study imidacloprid formulation containing 9.7% active ingredient, <2 % surfactant, and the co-solvent, *N*-methyl pyrrolidone caused drowsiness, disorientation, dizziness, oral and gastroesophageal erosions, hemorrhagic gastritis, productive cough, fever, leukocytosis, and hyperglycemia. In fact, moderate to high dose imidacloprid in animals causes central nervous system activation similar to nicotine, including tremors, impaired pupillary function, and hypothermia, however, the causal role for the toxicity is still unclear (Wu et al. 2001). Similar observation was mentioned that moderate to relatively high-dose imidacloprid in animals causes central nervous system activation similar to nicotine, including tremors, impaired papillary function, and hypothermia, it is more likely that the formulation ingredients caused most of the clinical symptoms including central nervous system depression and gastrointestinal irritation (Shadnia and Moghaddam 2008).

Two cases of acute poisoning with an insecticide formulation containing acetamiprid has been reported, exposed patients experienced severe nausea and vomiting, muscle weakness, hypothermia, convulsions, and clinical manifestations including tachycardia, hypotension, electrocardiogram changes, hypoxia, with the higher serum concentration of acetamiprid (Imamura et al. 2010). Similarly, another case study, Northeast China suggested toxic pesticides were responsible for comprehensive fatality (38.7%). Methomyl and fluoroacetamide were most commonly detected in the samples (Zhang et al. 2013).

In an autopsy case study of unnatural deaths in Northwest India, aluminium phosphide was found to be the most common suicidal poison, causing 68.4 % of total deaths due to poisoning between 1992 and 2000 (Chopra et al. 1986). Another case study revealed poisoning of 208 cases of death due to fatal poisoning of aluminium phosphide during the span of one year, January 2007 to December 2007. Studies decoded the mechanism of action of the fumigant in different animals described non-competitive binding of cytochrome oxidase by phosphine leading to valence change in the heme component of haemoglobin. However, other school of thoughts suggested inhibition of catalase, resulting to accumulation of hydrogen peroxide (Bogle et al. 2006).

Table 1: Case studies reported on clinical toxicology of pesticides

Year	Pesticide	Reported country	Patients	Fatality from acute poisoning	Symptoms	Pesticide content	Reference
2002-2007	Glyphosate-containing herbicides	Sri Lanka	601 patients	27% were asymptomatic, 63.7% had minor poisoning, and 5.5% had moderate to severe poisoning. 19 deaths with a median time to death of 20 h)	Gastrointestinal symptoms, respiratory distress, hypotension, altered level of consciousness, and oliguria	high plasma glyphosate concentrations on admission (>734 µg/mL)	Roberts et al. 2020
2002 -2009	Glyphosate-containing herbicides	France	13 cases	-	Oropharyngeal ulceration, nausea and vomiting, respiratory distress, cardiac arrhythmia, hyper-kaleamia, impaired renal function, hepatic toxicity and altered consciousness	Blood glyphosate concentrations had a mean value of 61 mg/L (range 0.6–150 mg/L) and 4146 mg/L (range 690–7480 mg/L)	Zouaoui et al. 2013
2001 - 2007	Endosulfan poisoning	China	52 cases	30.7% of the 52 patients died, and 48 patients experienced seizures.	Hypotension, rhabdomyolysis, hepatic toxicity, and hypotension.	Amount ingested could be greater than 35 g, caused mortality	Moon and Chun 2009
January to December 2005	Endosulfan poisoning	Singapore	23 cases	Nausea and vomiting in 17 patients (73.9 %), seizures 21.7 %, and dizziness 4.3 %	Nausea, vomiting, diarrhoea, stomach ache, respiratory distress, seizures, pulmonary oedema, agitation, headaches, dizziness etc.	-	Karatas et al. (2006)

Year	Pesticide	Reported country	Patients	Fatality from acute poisoning	Symptoms	Pesticide content	Reference
January 2001 to May 2003	Organophosphate pesticides (65%) and aluminium phosphide (15%)	India	325 patients	15% patients died	-	-	Singh and Unnikrishnan 2006
July 2013 to June 2014	Organophosphate pesticides	India	133 patients	36.1% were stable after gastric lavage	13.5% Patients showed episodic convulsions, derangement in renal function	-	Banday et al. 2015
1997- 2002	Monocrotophos and endosulfan	India	8040 patients	Fatality 22.6%, 1819 patient died	-	-	Srinivas Rao et al. 2005
2019	Monocrotophos	India	1 patient	Dead	Vomiting with breathlessness		Sharma and Kumar, 2019
July, 2008 -December 2015	Organophosphate pesticides	China	335 cases	-	Cerebral edema, liver injury, kidney injury, myocardial injury, gastrointestinal hemorrhage and acute pancreatitis	-	Zhang et al. (2017)
January 1985- December 1985	Aluminium phosphide	India	Sixteen patients		Profuse vomiting, pain in the upper abdomen and shock		Chopra et al. 1986

Toxicities of Chemical Pesticides as Potential Endocrine Disruptors

Many chemicals that have been identified as potential endocrine disruptors, include pesticides. About 105 substances have been identified in this respect, of these, 46% are insecticides, 21% herbicides and 31% fungicides. Case studies indicated that thyroid hormone production can be inhibited by some ten pesticides (amitrole, cyhalothrin, fipronil, ioxynil, maneb, mancozeb, pentachloronitro-benzene, prodiamine, pyrimethanil, thiazopyr, ziram, zineb (Sugiyama et al. 2005; Leghait et al. 2009). Besides, effects linked to endocrine disruption have been largely noticed in invertebrates (Gooding et al. 2003), reptiles (Crain et al. 1997), fish (Purdom et al. 1994), birds (Vos et al. 2000) and mammals (Oskam et al. 2003). A case study on *Daphnia magna* has shown that endosulfan sulphate disrupts the ecdysteroidal system and juvenile hormone activity of crustaceans (Palma et al. 2009). Influence of linuron on reproductive hormone production has been reported in rats (Wilson et al. 2009).

Epidemiological studies concluded that pesticide exposure affect spermatogenesis leading to poor semen quality and reduced male fertility, an increasing number of epidemiological studies linked environmental exposure to pesticides and hormone-dependent cancer risks. A case report on fat samples from women with breast cancer revealed elevated concentrations of PCBs, DDE, and DDT (Falck et al. 1992). Another epidemiological case studies performed in Spain between 1999 and 2009 shows that among 2,661 cases of breast cancer patients, 2,173 (81%) were associated with pesticide contamination (Parron et al. 2010).

Toxicities of Chemical Pesticides on Animals

Case report described the spraying of coca (*Erythroxylum coca*) with glyphosate (coca mixture, a combination of formulated glyphosate, Glyphos, and an adjuvant, Cosmo-Flux) in Colombia has raised concerns about possible impacts on amphibians. Mortality at 96 h in the control microcosms was between 0 and 16% and LC_{50} values were between 8.9 and 10.9 kg glyphosate a.i./ha. Mortality $>LC_{50}$ was only observed in the tested species when the application rate was >2-fold the normal application rate (Bernal et al. 2009). Contrastingly, another report suggested no significant acute toxicity of the glyphosate end-use formulation Roundup Original® to four North American amphibian species (*Rana clamitans, R. pipiens, R. sylvatica*, and *Bufo americanus*) and the toxicity of glyphosate technical, the polyethoxylatedtallowamine surfactant (POEA) commonly used in glyphosate-based herbicides, and five newer glyphosate formulations to *R. clamitans*. For *R. clamitans*, acute toxicity values in order of decreasing toxicity were POEA > Roundup Original > Roundup Transorb®>Glyfos AU®. However, relevant concentrations of POEA or glyphosate formulations containing POEA showed decreased snout-vent length at metamorphosis and increased time to metamorphosis, tail damage, and gonadal abnormalities. These effects may be caused, in some part, by disruption of hormone signaling, because thyroid hormone receptor β-m-RNA transcript levels were elevated (Howe et al. 2004).

The major mechanism of toxicity of OPs is the inhibition of Acetyl cholinesterase (AChE), resulting in a net accumulation of Acetylcholine (ACh) and increased stimulation of cholinergic receptors. In mammals, excessive stimulation of these cholinergic receptors in the central and peripheral nervous systems results muscarinic-receptor induced effects (excessive secretions, miosis, bradycardia) and nicotinic-receptor-induced effects (muscle tremors, convulsions, complete muscle paralysis). A case study in California, USA showed toxicity of phorate in a group of 300 Holstein cattle, a large number of cattle developed tremors, diarrhea, weakness, and paralysis. A total of 159, died within 24 h (Puschner et al. 2013).

An interesting study was carried out to assess the levels of atrazine, dimethoate, and dichlorodiphenyltrichloroethane on freshwater fish in Chiredzi, Zimbabwe revealed higher concentration of dichlorodiphenyltrichloroethane in water and fish muscle tissue at respective concentrations of 131.3 μg/l and 171.7 μg/kg, while concentrations of atrazine was 6.15 μg/l and 142.0 μg/kg in water and fish muscle tissue, respectively. The atrazine and DDT concentrations in water samples were above the limits permissible by the World Health Organization in drinking water. The pesticide in water were about three times higher than those in fish samples while significantly higher ($p < 0.05$) concentrations of atrazine (23-fold) were observed in fish samples compared to water. Levels of DDT and its metabolites in fish tissues were also higher than those in water samples (Basopo et al. 2020).

Another case study revealed the effect of glyphosate-based herbicide on aquatic organisms. Glyphosate has been widely used against terrestrial weeds, causes toxicity in plants include decreases in concentration of the aromatic amino acids, tryptophan, phenylalanine, and tyrosine, as well as decreased synthesis of protein, indole acetic acid and chlorophyll. Apart from their use in terrestrial environments, glyphosate-based formulas are also, however rarely, applied in order to control aquatic weeds, particularly invasive species e.g. common reed (*Phragmites australis*). Glyphosate was detected using gas chromatography mass spectrometry (GC-MS) in the water samples collected from the bathing area at a mean concentration of 0.09 mg dm^{-3}. Significantly lower numbers of Chironomidae (by 41%), Oligochaeta (by 43%), Vivipariae (by 75%), Hirudinae (by 75%), *Asellus aquaticus* (by 77%), *Gamma ruspulex* (by 38%) and *Dreissena polymorpha* (by 42%) were found at the glyphosate-treated site. The ranges of glyphosate concentration in the tissues of sampled macro-invertebrates and *P. australis* organs were 7.3-10.2 μg kg^{-1} and 16.2-24.7 μg kg^{-1}, respectively (Rzymski et al. 2013).

Another instance reported five dead or debilitated bald eagles (*Haliaeetus leucocephalus*) and a red-tailed hawk (*Buteo jamaicensis*) from British Columbia (BC), Canada tested positive for residues of the organophosphorus insecticide, phorate (Elliott et al. 1997). Other cases of pesticide poisoning of wild birds diagnosed at the National Veterinary Research and Quarantine Service, Korea, where forty-one mortality events (759 birds) of 87 incidents (2,464 birds) were found affected by poisoning of six organophosphates or carbamates pesticides. Phosphamidon was most frequently identified as the cause of poisoning, accounting for 23 mortality events, besides, other pesticides identified as poisons for birds were monocrotophus, fenthion, parathion, EPN, and diazinon, carbofuran (Kwon et al. 2004). Later, serious threat of carbofuran has been reported in western Kenya, where, uncounted dead birds, *Quelea* species found in cereal fields. On investigation, it was observed that a large proportion of individuals of their populations were exposed to Furadan (Odino 2011). Similar cases were reported from January 2014 to October 2020, which confirmed pesticide poisoning substances in 503 samples of wildlife and domestic animals in Portugal. Toxicology results from domestic species (dog, cat, sheep, cows, and horses), wildlife species (red foxes, birds of prey, lynx, and wild boar), molluscicides, carbamates, rodenticides, strychnine and organophosphates (Grilo et al. 2020).

Effect of Neonicotinoid Insecticides to Bees

Pollinating insects, such as the honey bee, are mainly exposed to chemicals when visiting melliferous plants. Neonicotinoid insecticides were recently implicated by beekeepers who reported that hives placed near cropped plants, originated from seeds dressed with insecticide, showed high levels of damage due to a progressive decrease in the hive populations, until the complete loss of the colonies (Maus et al. 2003). The risk that systemic neonicotinoid insecticides induce for honey bees started in France with the use of Gaucho (active ingredient: imidacloprid) on sunflower (Maxima and van der Sluijs 2007). Generally, neonicotinoids are more toxic via oral route than contact mode. The difference between the oral and contact toxicity may be due to the weak hydrophobicity of the neonicotinoidsyielding a low penetration through the insect cuticle. Three species of bees, *Apis mellifera*, *Megachile rotundata* and *Nomiam elanderi*, were found susceptible to imidacloprid (24-h LD_{50} 0.04 μg/bee) (Stark et al. 1995). Similar results were obtained for Admire and Provado that are two commercial formulations of imidacloprid (Devillers et al. 2003). The behavioral effects of neonicotinoid insecticides were largely investigated showed that foragers when collecting nectar and pollen were exposed to low doses of neonicotinoid insecticides during their foraging trips, which induced behavioral effects and subsequently no homing return to hive (Blacquiere et al. 2012).

A case study on the toxicity of dinotefuran, a neonicotinoid insecticide has been reported in Wilsonville, Oregon, USA which killed a large number of bumble bees. It was estimated that approximately 45,830 to 107,470 bumble bees from 289 to 596 colonies were killed with the chemical pesticide. Chromatographic analysis revealed concentration of dinotefuran in the samples flower was 7.4 ppm, which exceed 737% of the LC_{50} of beneficial pollinator, honey bee. The dead bumble bees were tested for dinotefuran concentration which showed 0.92 ppm, far more than the maximum LC_{50} (0.884 ppm) of *A. melifera* (Hatfield et al. 2021). Another study reported in Romania, where exposure of honey bees with neonicotinoids was estimated. In total, a set of fifty samples was collected from fields, located in different areas of intensive agriculture were analysed for five neonicotinoids which revealed 48% of the total samples contained one or more detected or quantified neonicotinoid residues (Căuia et al. 2020).

Instances of Environmental Toxicities of Chemical Pesticides

Despite the presence of rules and regulations, it has been observed that pesticides are not used in an appropriate manner. Much of the portion of chemical pesticides goes to wastage during their use. Pesticides are used in various types of pests control; remain a big source of air, water and soil pollution, which may negatively affect human health and the living organisms in the environment. Environmental impacts of pesticide use were commonly estimated through variables such as pounds of active ingredient applied or expenditure on pesticides. The disadvantage is that both these measures assume environmental damage is directly correlated with the quantity of pesticide used, regardless of the specific chemical formulation. The increased availability of low-dosage alternatives lend credence to the argument that weight and volume measures are not adequate proxies for assessing pesticide risk. Cornell University's Environmental Risk Analysis Program has identified eight of the indicators widely used worldwide: Environmental Potential Risk Indicator for Pesticides (EPRIP), Environmental Yardstick for Pesticides (EYP), Survey of National Pesticide Risk Indicators (SYNOPS), System for Predicting the Environmental Impact of Pesticides (SyPEP), Pesticide

Environmental Risk Indicator (PERI), Environmental Impact Quotient (EIQ), Chemical Hazard Evaluation for Management Strategies (CHEMS1), and Multi-Attribute Toxicity Factor (MATF). The first four indicators are referred to as predicted environmental concentration (PEC) indicators, and the later four constitute ranking indicators (Sande et al. 2011).

Negative effects of pesticides on the environment and the farmers awareness in Saudi Arabia has been described by Al-Zaidi et al. (2011). Another case study on assessment of hazards from methyl bromide and the proposed alternative fumigants to workers, consumers, beneficial arthropods, birds, fish, and bees in Florida, USA suggested the highest relative risks category under field workers and beneficial arthropods and fish and consumers the least risks (Sande et al. 2011). Similar kind of study was also conducted in Nepal to evaluate vegetable growers' knowledge on pesticide safety and pest management practices. Unfortunately, most of the farmer (>90%) did not know much about the harmful effects of pesticide residues nor practiced proper pesticide disposal methods (Rijal et al. 2018). A case study conducted the health risk associated with chemical pesticide contaminants in the drinking water sources of Dalian in China, revealed relatively higher concentration of atrazine and acetochlorat ng L^{-1} levels. Additionally, atrazine, acetochlor, hexachlorobenzene, p,p'-DDE, and p,p'-DDD were detected in the sediment/soil samples at ng g-1 levels. However, hexachlorobenzene and arsenic were identified as the main contributors to human carcinogenic risks, which were calculated at the high level of 10^{-4} (Dong et al. 2020).

Why Biopetsicides?

Biopesticide is gaining interest because of its advantages associated with the environmental safety, target-specificity, efficacy, biodegradability and suitability in the integrated pest management (IPM) programs. Thus, biopesticide is one of the promising alternatives to manage environmental pollutions. Though potential application of biopesticides in environmental safety is well known, it has gained interest in view of the growing demands for organic food. Although use of agrochemicals is indispensable to meet the ever growing demands of food, feed and fodder, opportunities do exist in selected crops and niche areas where biopesticides can be used as a component of IPM. The interest in biopesticides is based on the advantages associated with the products which are (i) inherently less harmful and environmentally safe, (ii) target-specific, (iii) often effective in very small quantity, (iv) naturally and quickly decomposable and, (v) usable as a component of IPM.

Biopesticides are very effective in the agricultural pest control without causing serious harm to ecological chain or worsening environmental pollution. The research and development of practical applications in the field of biopesticides greatly mitigate environmental pollution caused by chemical pesticide residues and promotes sustainable development of agriculture. Since the advent of biopesticides, a large number of products have been released, several of which have already played dominant roles in the market. The development of biopesticides stimulates modernization of agriculture and will, without doubt, gradually replace chemical pesticides. Many biopesticides are ideal substitutes for their traditional chemical counterparts in pollution-free agricultural production. Research in production, formulation and delivery may greatly assist in commercialization of biopesticides. More research is needed towards integrating biological agents into production system, improving capability of developing countries to manufacture and use biopesticides. At the same time, it is also required to encourage public funded programmes, commercial investors and pesticide companies to take up biopesticide enterprises.

Conclusions

Impact of synthetic pesticides, due in particular to an excessive use (including environmental pollution and implications to human health) have led to modifications in agricultural practices and various national and international regulations limiting their use. Further limitations and/or bans often encourage to find alternative solutions that are safer and non-toxic to the environment and humans. Most of the countries have amended their policies to minimize the use of chemical pesticides and promote the use of biopesticides. Policy measures need to be strengthened in order to reduce excessive use of chemical pesticides and to promote the use of biopesticides. Better understanding of the mode of action of biopesticides, their effects and regulatory issues that arise in their adoption may help further to raise their profile among the public, policy-makers and hence enable them to realize their contributions to sustainability. The interest in organic farming and pesticide residue free agricultural produce would certainly warrant increased adoption of biopesticides by the farmers. Increasing concerns over environmental and health safety across the world would certainly create awareness among the farmers, manufacturers, policy makers and consumers to accept safer biopesticides for suitable pest management options.

References

Akhgari M, Amini SN, Iravani FS (2018) Forensic toxicology perspectives of methadoneassociated deaths in Tehran, Iran, a 7 year overview. Basic Clin Pharmacol Toxicol 122(4):436-441.

Al-Zaidi AA, Elhag EA, Al-Otaibi SH, Baig MB (2011) Negative effects of pesticides on the environment and the farmers awareness in Saudi Arabia: A case study. J Animal Plant Sci 21(3):605-611.

Banday TH, Tathineni B, Desai MS, Naik V (2015) Predictors of morbidity and mortality in organophosphorus poisoning: a case study in rural hospital in Karnataka, India. North Am J Med Sci 7(6):259.

Basopo N, Muzvidziwa A (2020) Assessment of the effects of atrazine, dichlorodiphenyltrichloroethane, and dimethoate on freshwater fish (*Oreochromismoss ambicus*): A case study of the a2 farmlands in chiredzi, in the southeastern part of Zimbabwe. Environ Sci Pollut Res 27(1):579-586.

Bernal MH, Solomon KR, Carrasquilla G (2009) Toxicity of formulated glyphosate (Glyphos) and Cosmo-Flux to larval and juvenile Colombian frogs 2. Field and laboratory microcosm acute toxicity. J Toxicol Environ Health, Part A 72(15-16):966-973.

Blacquiere T, Smagghe G, Van Gestel CA, Mommaerts V (2012) Neonicotinoids in bees: a review on concentrations, side-effects and risk assessment. Ecotoxicol 21(4):973-992.

Bodwal J, Chauhan M, Behera C, Jitendra K (2019) Fatal monocrotophos poisoning by cutaneous absorption while sleeping: A remarkable case study. Med Leg J 87(3):144-150.

Bogle RG, Theron P, Brooks P, Dargan PI, Redhead J (2006) Aluminium phosphide poisoning. Emerg Med J 23(1):e3.

Căuia E, Siceanu A, Visan GO, Căuia D, Colta T, Spulber RA (2020) Monitoring the field-realistic exposure of honeybee colonies to neonicotinoids by an integrative approach: A case study in Romania. Diversity 12(1):24.

Chopra JS, Kalra OP, Malik VS, Sharma R, Chandna A (1986) Aluminium phosphide poisoning: a prospective study of 16 cases in one year. Postgrad Med J 62(734):1113-1115.

Chugh SN, Dhawan R, Agrawal N, Mahajan SK (1998) Endosulfan poisoning in Northern India: a report of 18 cases. International J Clin Pharmacol Ther 36(9):474-477.

Crain DA, Guillette LJ, Pickford DB, Rooney A (1997) Alterations in steroidogenesis in alligators (*Alligator mississippiensis*) exposed naturally and experimentally to environmental contaminants. Environ Health Perspect 105:528-533.

Devillers J, Decourtye A, Budzinski H (2003) Comparative toxicity and hazards of pesticides to *Apis* and non-*Apis* bees. A chemometrical study, SAR QSAR. Environ Res 14:389-403.

Dong W, Zhang Y, Quan X (2020) Health risk assessment of heavy metals and pesticides: a case study in the main drinking water source in Dalian, China. Chemosphere 242:125113.

Elliott JE, Wilson LK, Langelier KM, Mineau P, Sinclair PH (1997) Secondary poisoning of birds of prey by the organophosphorus insecticide, phorate. Ecotoxicol 6(4):219-231.

Falck F, Ricci A, Wolff MS, Godbold J, Deckers P (1992) Pesticides and polichlorinated biphenyl residues in human breast lipids and their relation to breast cancer. Ach Environ Occup Health 47:143-146.

Gooding MP, Wilson VS, Folmar LC, Marcovich DT, LeBlanc GA (2003) The biocide tributyltin reduces the accumulation of testosterone as fatty acid esters in the mud snail (*Ilyanassa obsoleta*). Environ Health Perspect 111:426-230.

Gress S, Lemoine S, Séralini GE, Puddu PE (2015) Glyphosate-based herbicides potently affect cardiovascular system in mammals: review of the literature. Cardiovas Toxicol 15(2):117-126.

Grilo A, Moreira A, Moreira B, Belas A, São Braz B (2020) Epidemiological study of pesticide poisoning in domestic animals and wildlife in Portugal: 2014-2020. Front Vet Sci 7:1181.

Gunnell D, Eddleston M, Phillips MR, Konradsen F (2007) The global distribution of fatal pesticide self-poisoning: systematic review. BMC Public Health 7(1):1-15.

Hatfield RG, Strange JP, Koch JB, Jepsen S, Stapleton I (2021) Neonicotinoid pesticides cause mass fatalities of native bumble bees: A case study from Wilsonville, Oregon, United States. Environ Entomol. doi.org/10.1093/ee/nvab059

Hori Y, Fujisawa M, Shimada K, Hirose Y (2003) Determination of the herbicide glyphosate and its metabolite in biological specimens by gas chromatography-mass spectrometry. A case of poisoning by Roundup® herbicide. J Anal Toxicol 27(3):162-166.

Howe CM, Berrill M, Pauli BD, Helbing CC, Werry K, Veldhoen N (2004) Toxicity of glyphosatebased pesticides to four North American frog species. Environ Toxicol Chem 23(8):1928-1938.

Huang NC, Lin SL, Chou CH, Hung YM, Chung HM, Huang ST (2006) Fatal ventricular fibrillation in a patient with acute imidacloprid poisoning. Am J Emerg Med 24(7):883-885.

Imamura T, Yanagawa Y, Nishikawa K, Matsumoto N, Sakamoto T (2010) Two cases of acute poisoning with acetamiprid in humans. Clin Toxicol 48(8):851-853.

Karatas AD, Aygun D, Baydin A (2006) Characteristics of endosulfan poisoning: a study of 23 cases. Singapore Medical Journal 47(12):1030-1032.

Kır MZ, Öztürk G, Gürler M, Karaarslan B, Erden G, Karapirli M, Akyol Ö (2013) Pesticide poisoning cases in Ankara and nearby cities in Turkey: an 11-year retrospective analysis. J Forensic Leg Med 20(4):274-277.

Kumar A, Verma A, Kumar A (2013) Accidental human poisoning with a neonicotinoid insecticide, imidacloprid: A rare case report from rural India with a brief review of literature. Egyptian J Forensic Sci 3(4):123-126.

Kumar S (2013) The role of biopesticides in sustainably feeding the nine billion global populations. J Biofertil Biopestic 4:e114.

Kwon YK, Wee SH, Kim JH (2004) Pesticide poisoning events in wild birds in Korea from 1998 to 2002. J Wildl Dis 40(4):737-740.

Langrand J, Blanc-Brisset I, Boucaud-Maitre D, Puskarczyk E, Nisse P, Garnier R, Pulce C (2020) Increased severity associated with tallowamine in acute glyphosate poisoning. Clin Toxicol 58(3):201-203.

Lee HL, Chen KW, Chi CH, Huang JJ, Tsai LM (2000) Clinical presentations and prognostic factors of a glyphosate surfactant herbicide intoxication a review of 131 cases. Acad Emerg Med 7(8):906-910.

Leghait J, Gayrard V, Picard-Hagen N, Camp M, Perdu E, Toutain PL, Viguié C (2009) Fipronil-induced disruption of thyroid function in rats is mediated by increased total and free thyroxine clearances concomitantly to increased activity of hepatic enzymes. Toxicol 255:38-44.

Mancini F, Van Bruggen AH, Jiggins JL, Ambatipudi AC, Murphy H (2005) Acute pesticide poisoning among female and male cotton growers in India. Int J Occup Environ Health 11(3):221-232.

Maus C, Curé G, Schmuck R (2003) Safety of imidacloprid seed dressings to honey bees: a comprehensive overview and compilation of the current state of knowledge. Bull Insectology 56(1):51-57.

Maxima L, van der Sluijs JP (2007) Uncertainty: Cause or effect of stakeholders' debates?: Analysis of a case study: The risk for honeybees of the insecticide Gaucho. Sci Total Environ 376(1-3):1-17.

Moon JM, Chun BJ (2009) Acute endosulfan poisoning: a retrospective study. Hum Exp Toxicol 28(5):309-316.

Odino M (2011) Measuring the conservation threat to birds in Kenya from deliberate pesticide poisoning: a case study of suspected carbofuran poisoning using Furadan in Bunyala Rice Irrigation Scheme. Carbofuran and Wildlife Poisoning: Global Perspectives and Forensic Approaches 53-70.

Oskam IC, Ropstad E, Dahl E, Lie E, Derocher AE, Wiig O, Larsen S, Wiger R, Skaare JU (2003) Organochlorines affect the major androgenic hormone; testosterone; in male polar bears (*Ursus maritimus*) at Svalbard. J Toxicol Environ Health A 66:2119-2139.

Palma P, Palma VL, Matos C, Fernandes RM, Bohn A, Soares AMVM, Barbosa IR (2009) Effects of atrazine and endosulfan sulphate on the ecdysteroid system of *Daphnia magna*. Chemosphere 74:676-681.

Parron T, Alarcon R, Requena MDM, Hernandez A (2010) Increased breast cancer risk in women with environmental exposure to pesticides. Toxicol Lett 196:S180.

Patel AB, Dewan A, Kaji BC (2012) Monocrotophos poisoning through contaminated millet flour. Arh za Hig Rada Toksikol 63(3):377.

Peillex C, Pelletier M (2020) The impact and toxicity of glyphosate and glyphosate-based herbicides on health and immunity. J Immunotoxicol 17(1):163-174.

Purdom CE, Hardiman, PA, Bye VJ, Eno NC, Tyler CR, Sumpter JP (1994) Estrogenic effects of effluents from sewage treatment works. Chem Ecol 8:275-285.

Puschner B, Gallego S, Tor E, Wilson D, Holstege D, Galey F (2013) The diagnostic approach and public health implications of phorate poisoning in a California Dairy Herd. J Clin Toxicol S13. doi:10.4172/2161-0495.S13-001

Rijal JP, Regmi R, Ghimire R, Puri KD, Gyawaly S, Poudel S (2018) Farmers' knowledge on pesticide safety and pest management practices: A case study of vegetable growers in Chitwan, Nepal. Agriculture 8(1):16.

Roberts DM, Nick AB, Fahim M, Michael E, Daniel A, Goldstein AM, Marian SB, Andrew HD (2010) A prospective observational study of the clinical toxicology of glyphosate-containing herbicides in adults with acute self-poisoning. Clin Toxicol 48(2):129-136.

Rzymski P, Klimaszyk P, Kubacki T, Poniedzialek B (2013) The effect of glyphosate-based herbicide on aquatic organisms-a case study. Limnol Rev 13(4):215.

Sande D, Mullen J, Wetzstein M, Houston J (2011) Environmental impacts from pesticide use: a case study of soil fumigation in Florida tomato production. Int J Environmental Res Public Health 8(12):4649-4661.

Shadnia S, Moghaddam HH (2008) Fatal intoxication with imidacloprid insecticide. Am Emerg Med 26(5):634.e1-4. doi: 10.1016/j.ajem.2007.09.024.

Sharma R, Kumar J (2019) Fatal poisoning of monocrotophos pesticide by subcutaneous absorption: a case study. Malaysian J Forensic Sci 9(1):15-20.

Singh B, Unnikrishnan B (2006) A profile of acute poisoning at Mangalore (South India). J Clin Forensic Med 13(3):112-116.

Sribanditmongkol P, Jutavijittum P, Pongraveevongsa P, Wunnapuk K, Durongkadech P (2012) Pathological and toxicological findings in glyphosate-surfactant herbicide fatality: a case report. The American J Forensic Med Pathol 33(3):234-237.

Srinivas Rao CH, Venkateswarlu V, Surender T, Eddleston M, Buckley NA (2005) Pesticide poisoning in south India: opportunities for prevention and improved medical management. Trop Med Int Health 10(6):581-588.

Srivastava N, Harit G, Srivastava R (2009) A study of physico-chemical characteristics of lakes around Jaipur, India. J Environ Biol 30(5):889.

Stark JD, Jepson PC, Mayer DF (1995) Limitation to use of topical toxicity data for prediction of pesticide side effect in the field. J Econ Entomol 88:1081-1088.

Sugiyama S, Shimada N, Miyoshi H, Yamauchi K (2005) Detection of thyroid systemdisrupting chemicals using *in vitro* and *in vivo* screening assays in *Xeno puslaevis*. Toxicol Sci 88:367-374.

Talbot AR, Mon-Han S, Jinn-Sheng H, Shu-Fen Y, Tein-Shong G, Shur-Hueih W, Chao-Liang C, Thomas RS (1991) Acute poisoning with a glyphosate-surfactant herbicide ('Roundup'): a review of 93 cases. Hum Exp Toxicol 10(1):1-8.

Teixeira H, Proença P, Alvarenga M, Oliveira M, Marques EP, Vieira DN (2004) Pesticide intoxications in the Centre of Portugal: three years analysis. Forensic Sci Int 143(2-3):199-204.

Tominack RL, Yang GY, Tsai WJ, Chung HM, Deng JF (1991) Taiwan National Poison Center survey of glyphosate-surfactant herbicide ingestions. J Clin Toxicol 29(1):91-109.

UNDESA (United Nations Department of Economic and Social Affairs) (2009) World population prospects: The 2008 revision, highlights, working Paper No. ESA/P/WP.210. New York, UN.

Vos JG, Dybing E, Greim HA, Ladefoged O, Lambré C, Tarazona JV, Brandt I, Vethaak AD (2000) Health effects of endocrine-disrupting chemicals on wildlife; with special reference to the European sitation. Crit Rev Toxicol 30:71-133.

Wilson VS, Lambright CR, Furr JR, Howdeshell KL, Gray LEJ (2009) The herbicide linuron reduces testosterone production from the fatal rat testis during both in utero and *in vitro* exposures. Toxicol Lett 186:73-77.

Wu IW, Lin JL, Cheng ET (2001) Acute poisoning with the neonicotinoid insecticide imidacloprid in N-methyl pyrrolidone. J Clin Toxicol 39(6):617-621.

Yavuz Y, Yurumez Y, Kücüker H, Ela Y, Yüksel S (2007) Two cases of acute endosulfan toxicity. Clin Toxicol 45(5):530-532.

Zhang D, Zhang J, Zuo Z, Liao L (2013) A retrospective analysis of data from toxic substance-related cases in Northeast China (Heilongjiang) between 2000 and 2010. Forensic Sci Int 231(1-3):172-177.

Zhang W, Huang C, Jiang Y, Bai L, Zhang X (2017) Comparative study of the incidence of early complications among the patients with six different kinds of acute organophosphorus pesticide poisoning. Chinese J Emerg Med 26(11):1247-1251.

Zouaoui K, Dulaurent S, Gaulier JM, Moesch C, Lachâtre G (2013) Determination of glyphosate and AMPA in blood and urine from humans: About 13 cases of acute intoxication. Forensic Sci Int 226(1-3):20-25.

2

Insecticide Resistance and Case Histories

Abstract

Insecticides play a crucial role in the management of insect pests in order to reduce yield losses caused to high value and cash crops not only in India but also in the world. In the recent past few newer insecticide molecules have come in place to overcome hazardous issues with chemical pesticides. However, their usage is very limited and such molecules are not available for high value cash crops in case of pest outbreaks, resurgence etc. Insecticide resistance has got a long history wherein many insect pests have developed significant level of resistance making insecticides ineffective. Insecticide resistance in India started with development of resistance in Singhara beetle against DDT. Various resistance mechanisms adopted by insects to combat toxic effects are metabolic, altered target site sensitivity (mutations), reduced penetration, behavioural resistance etc. Pink bollworm developed resistance to Bollgard II in the country caused huge economic losses to cotton crop. Even recent outbreak of whiteflies in cotton in Punjab has also devastated the entire crop due to development of resistance. Although insecticide resistance management strategies have come in place for few pests, the insecticide resistance in insects in major crops made the various stake holders to think seriously for alternative pest control strategies to safe guard the crops from ravaging insect pests. Green pesticides, biopesticides and pesticides/ insecticides of biological origin have got greater attention as consumers are health conscious and demanding for organic products. Hence, there is paradigm shift in pest management strategies from chemical control to microbial control in India as well as in the world.

Keywords: Pest, Resistance, Insecticide, Mechanism, Biopesticides

Introduction

Insecticides are one of the key control measures to combat the insect pests for sustainable agricultural production in the world. Synthetic insecticides have been only strategy to control the resurgent and resistant insect population in of high value crops not only in India but also in the world. The advent of synthetic insecticides in the mid-20th century made the control of insect and other arthropod pests easy and much more effective, and such chemicals remain essential in modern agriculture despite their environmental issues. By preventing crop losses, raising the quality of produce, and lowering the cost of farming, modern insecticides increased crop yields by as much as 50% in some regions of the world. More than half of our crops would be lost to pests and diseases if pesticides are not employed. Between 26 and 40% of the world's potential crop production is lost annually because of weeds, pests and diseases (OECD-FAO 2012). Without crop protection, these losses could easily double. Insecticides enable farmers to produce safe, quality foods at affordable prices with abundance of nutritious, all-year-round foods, which are necessary for human health. Fruits and vegetables, which provide essential nutrients, are more abundant and affordable. Grains, milk and proteins, which are vital to childhood development, are more widely available because of lower costs to produce food and animal feed. Production of major crops has more than tripled since 1960, which was mostly due to pesticides (FAOSTAT) as in case of rice which feeds almost half the people on our planet doubled in production while the amount of wheat has increased nearly 160%. Insecticides have also been important in improving the health of both humans and domestic animals; malaria, yellow fever, and typhus, among other infectious diseases, have been greatly reduced in many areas of the world through their use. Pesticides include insecticides that are mainly used in agriculture or in public health protection programs in order to protect plants from pests, weeds or diseases, and humans from vector-borne diseases, such as malaria, dengue fever, and schistosomiasis (Alewu and Nosiri 2011). Besides, insecticides are being extensively used in sports fields, building bottoms, lawn development, public urban green areas etc to prevent unwanted insect pests such as termites (Hoffman et al. 2000; Nicolopoulou-Stamati et al. 2016). Despite tremendous benefits and advantages offered by insecticides to mankind, continuous use of insecticides over a longer period of time resulted in development of resistance, resurgence, residue and environmental issues, health hazards etc. Insecticide resistance become one of the major concerns in agriculture, public health sector and other fields in India as well as in the world.

Insecticide and Insecticide Resistance

Insecticide is an agent that destroys insects as well as other small pests (such as mites or nematodes). While WHO panel of experts defined insecticide resistance as 'the development of an ability of a strain of insects to tolerate doses of toxicants which would prove lethal to the majority of individuals in a normal population of the same species' (Guedes 2017). Resistance in insects is usually a complex phenomenon with more than one mechanism operating simultaneously within the same insect strain (Oppenoorth and Welling 1976). The resistant phenotype of an insect that survives a dose of insecticide that would normally have killed it, is relatively monitored with direct insecticide bioassays. Pest control subjects the population to Darwinian selection and survival of the fittest and it attempts to kill the tolerant individuals lead to ever increasing doses and eventually resistant pest populations. As a result, the most difficult problems raised due to insensitive biochemical target conferring cross resistance to one or more classes of compounds formerly effective at that site. Insecticides are being widely used to control insect pests across the world which leads to high selection pressure on target insect over a period of time. Insecticides provide very good control of insects initially but over a period of time insects develop resistance by various mechanisms such as morphological, behavioural, ecological, environmental biochemical, genetical, molecular adaptations.

Importance of Insecticide Resistance and its Monitoring

Since the 1950s insecticide resistance has come into prominence around the globe as a key factor impacting the use and efficacy of a wide range of existing and new compounds for the control of insect and mite crop pests as well as vectors of human diseases. Insecticide resistance is also an important driver in the search for new insecticides, especially those with new modes of action. Within the crop protection industry, insecticide resistance was recognized as a concern as early as the late 1950s to early 1960s. The early industry response most often involved simply finding and using a different insecticide. Frequently the replacement products were in the same class of chemistry since there were few distinct classes of insecticides available during time. However, in some instances recommendations from industry, scientists included specific resistance mitigating measures such as moderation of use, alternation (rotation) of insecticides from different classes, and incorporation of biological control measures.

Mechanisms of Insecticide Resistance

Insect resistance to insecticides has been found to be mediated by various mechanisms in four different ways (Krathi et al 2002; Ju et al 2021, Liu 2015, Auteri et al 2018) :

a) Metabolic resistance, due to an increased detoxification caused by the overexpression or conformational changes of the enzymes involved in the chemical insecticide metabolism, sequestration, and excretion. Cytochrome P450-monooxygenases, glutathione S-transferases, and carboxy/cholinesterases, microsomal mono-oxygenases, phosphotriester hydrolases, DDT-dehydrochlorinases are the main enzymes involved in this process.

b) Altered target site sensitivity/mutation, caused by a modification of the chemical insecticide site of action reducing or preventing insecticide binding at that site. Mutations in the voltage sensitive sodium channel (Vssc) gene are one of the most common causes of target-site resistance. Insensitive acetylcholinesterases, insensitive sodium channels, insensitive GABA (γ-amino butyric acid) receptor are few such examples.

c) Reduced penetration, due to modifications in the insect cuticle or digestive tract linings that limit the chemical insecticides absorption. However, the mechanism remains poorly understood, and its importance in Aedes species is yet to be confirmed.

d) Behavioural resistance, which includes modifications in insect behaviour that help to avoid the lethal effects of chemical insecticides. This is considered as a contributing factor that allows the insects to avoid the lethal dose of an insecticide.

Insects metabolize insecticides to non toxic or less toxic forms through a process called 'detoxification' and sometimes to more toxic intermediates, a process called 'activation'. Substances that are completely water soluble (polar), and those that are completely insoluble in either water or fats, are excreted unchanged. Most insecticides, which are water insoluble (apolar) or fat soluble (lipophilic), are metabolised to polar compounds through a primary enzymatic conversion, mediated through 1) Oxidases, 2) Hydrolases or 3) Glutathione-S-transferases, resulting in watersoluble products that are subsequently converted to water soluble conjugates through a secondary non synthetic reaction. These conjugates are finally excreted. Apolar substances are converted to less lipophilic or polar metabolites by two reactions (Phase I and Phase II) in insects and many other organisms.

Phase I reactions are mainly carried out by two major groups of enzymes, the oxidoreductases and hydrolases. The oxidoreductases comprise of the cytochrome P450 dependent superfamily of monooxygenases, which introduce oxygen into or remove electrons from their substrates. Carbonyl reductases, alcohol dehydrogenases and aldehyde dehydrogenases remove hydrogen from, or add to the target molecule. The hydrolases hydrolyse esters, amides, epoxides or glucuronides. Typically the Phase I reaction introduces a functional group in a series of steps in lipophilic molecules. Phase II reactions are mainly carried out by the transferases. Glutathione S-transferases conjugate the electrophilic substrates, while the acetyl transferases, sulfotransferases, acyl-CoA aminoacid N-methyl transferases and UDP-glucuronosyl transferases metabolise the nucleophilic substrates. Insecticide metabolism in insects has been found to be catalysed mainly by monooxygenases, hydrolases and gluthathione-S-transferases. Generally, in resistant insects, the enzymatic detoxification is believed to be so rapid that the toxic molecule does not reach its site of toxic action.

History of Insecticide Resistance

In India, insecticide resistance has been well documented by Mehrotra (1989) where Singhara beetle, *Galerucella birmanica* (1963), Tobacco caterpillar, *Spodoptera litura* (1965), Diamondback moth, *Plutella xylostella* (1968), Gram pod borer, *Heliothis armigera* (1986), aphids and jassids, *Empoasca kerri* (1986), *Lipaphis erysimi* (1986), *Aphis craccivora* (1986) have developed resistance to DDT, HCH, organophosphates (malathion and dimethoate), endosulfan etc. Subsequently, *Helicoverpa armigera* (1987) developed resistant to synthetic pyrethroids in cotton ecosystem where very high proportion of insecticides have been used before the introduction of Bt cotton. Insecticide resistance has been reported mostly in cotton ecosystem during the decade 1990-2000 which was the most difficult for cotton pest management due to excessive use of insecticides, especially synthetic pyrethroids that led to problems of high levels of resistance to pyrethroids and DDT in *Helicoverpa armigera* and *Spodoptera litura* in cotton and pulse growing regions of the country (Sekhar et al. 1996). Subsequent studies (Armes et al. 1996; Kranthi et al. 2002) showed that resistance to pyrethroids was ubiquitous and resistance in *H. armigera* to conventional insecticides such as methomyl, endosulfan and quinalphos was increasing in India. Due to unsatisfactory insect control on account of insecticide resistance, farmers were forced to spray repeatedly, most often with mixtures.

The outbreaks of whitefly during 1988 and recently on cotton in Punjab (2015) which destroyed 2/3 of cotton in India was due to indiscriminate use of pyrethroids, elimination of natural enemies, favourable temperatures (Sundaramurthy et al. 1992), presence of a wide range of hosts such as vegetables, pulses, throughout year helps the whiteflies to survive and proliferate (Kranthi 2015). Moreover, development of resistance in whiteflies to synthetic pyrethroids and insect growth regulators like pyriproxyfen was reported. Very recently, Naveen et al. (2017) Studied the level of insecticide resistance to selected organophosphates, pyrethroids, and neonicotinoids in seven Indian field populations of *Bemisia tabaci* genetic groups Asia-I, Asia-II-1, and Asia-II-7. Asia-I and Asia-II-1 populations were showing significant resistance wherein LC^{50} values were 7x for imidacloprid and thiamethoxam, 5x for monocrotophos and 3x for cypermethrin among the Asia-I, while, they were 7x for cypermethrin, 6x for deltamethrin and 5x for imidacloprid within the Asia-II-1 populations. A substantial increase in resistance ratios was observed in both the populations of Asia-I and Asia-II-1. It is evident that potential control failure was detected using probit analysis estimates for cypermethrin, deltamethrin, monocrotophos and imidacloprid in controlling whiteflies due to significant development of insecticide resistance. Insecticide resistance has been reported in public health in India in many insect vectors species.

Chinnababu Naik et al. (2018) reported the mean Resistance Ratio (RR) for *cry*1Ac against PBW was 47 during 2013 and the has increased to 1387 during 2017. A similar increasing trend was observed for *cry* 2Ab with a mean RR increase from 5.4 in 2013 to 4196 in 2017. Widespread infestation of pink bollworm in Bt cotton ranging between 40 – 95% accounting for estimated yield losses to the tune of 20-30 % have been reported from 16 major cotton growing districts of Maharashtra, a leading cotton producing state of Central India (Kranthi 2015). This was because of the evolution of resistance in PBW against bollgard II (*cry*1Ac & *cry*2Ab) of *Bt* cotton hybrid in India . The causes of resistance were insufficient refuge, extended growing season, lower expression levels of cry toxin in later crop growth stage etc lead to outbreakof PBW in Indian cotton (Fand et al. 2019). Experts primarily pointed that abundance of refuge varied among countries that might have played a key role in the striking differences in the incidence of the same pest species on the same crop and on the same toxins, without discounting the role of other differing factors like nature of hybrids and varieties, climate and production practice adopted in the three major cotton growing countries in the world. In addition, number of stored grain pests have also developed resistance to insecticides in India. Pesticide resistance in stored grain pests appeared comparatively later and first reported in flour beetle, *Tribolium castaneum*, in 1971 against DDT and malathion from Delhi followed by lindane and phosphine. *Sitophilus oryzae*, another serious pest of stored grain, becoming resistant to malathion originated from Kanpur in 1973 and also to lindane and phosphine. *Rhizopertha dominica, Ryzaephilus surinamensis, Dermestes maculatus* have also witnessed resistant to malathion, lindane and phosphine in various parts of the country (Mehrotra, 1989).

The pesticide resistance in India was first noticed in insect pests of public health importance and the concern about it led to an International Conference organised pointedly by the World Health Organisation and the Government of India in 1958 at New Delhi. Mosquitoes transmitting malaria and other vector diseases were the first to become resistant to pesticides. This was because of the large scale use of DDT in the National Malaria Control Programme / National Malaria Eradication Programme. The first report of DDT resistance in mosquitoes came in1952 from UP and Bombay in *Culex fatigans* a transmitter of filaria and has been reported to be resistant to both DDT and HCH in various parts of the country. The resistance in urban malaria transmitter, *Anopheles stephensi,* to DDT was reported first from Erode, Tamil Nadu in 1956. In fact *A. culicifacies* which accounts for more than 70% of the rural malaria is resistant throughout the country to one or the other pesticide used in Malaria Control Programme.

In the world, insecticide resistance has been reported in many occasions. Melander (1914) reported the first case of insecticide resistance to lime sulphur, an inorganic insecticide, in an orchard pest, the San Jose scale (*Quadraspidiotus perniciosus*) in the state of Washington. A treatment with lime sulphur killed all scales in one week in typical orchards, but 90 percent survived after two weeks in an orchard with resistant scales. Subsequently, the number of insecticide resistance cases grew exponentially following widespread use of DDT and other synthetic organic insecticides. Insects have evolved resistance to all types of insecticides including in-organics, DDT, cyclodienes, organophosphates, carbamates, pyrethroids, juvenile hormone analogs, chitin synthesis inhibitors, avermectins, neonicotinoids, and microbials. Since the first case of DDT resistance in 1947, the incidence of resistance has increased annually at an alarming rate. It has been estimated that there are at least 447 pesticide resistant arthropods species in the world (Callaghan, 1991). Insecticide resistance has also been developed by many insects to new insecticides with different mode of action like neoniconitoids. Resistance reported in thirteen orders of insects, yet more than 90 percent of the arthropod species with resistant populations are either Diptera (35 percent), Lepidoptera (15 percent), Coleoptera (14 percent), Hemiptera (14 percent), or mites (14 percent). The disproportionately high number of resistant Diptera reflects intense use of insecticides against mosquitoes that transmit disease, and agricultural pests account for 59 percent of harmful resistant species while medical and veterinary pests account for 41 percent. Statistical analyses suggest that for crop pests, resistance evolves most readily in those with an intermediate number of generations (four to ten) per year that feed either by chewing or by sucking on plant cell contents.

In 1990's, neonicotinoids includes imidacloprid, clothianidin, and thiamethoxam have been introduced into the global market as alternatives to organophosphates and carbamates to control sucking and other pests and they proved good for a while. But subsequently, neonicotinoids have proved the development of resistance in *Myzus persicae* and *Phorodon humuli*. The effects of imidacloprid on *Nilaparvata lugens*, tebufenozide on *Plutella xylostella* and *Spodoptera exigua*, thiamethoxam on *Bemisia tabaci*, trichlorphon on *Bactrocera dorsalis*, imidacloprid on *Spodoptera litura*, and emamectin benzoate on *Chrysoperla carnea* have been (Sahani and Pal 2021). The first report of neonicotinoid resistance was published in 1996, describing the low efficacy of imidacloprid against Spanish greenhouse populations of cotton whitefly. Later-generation, show stronger resistance (up to 17-fold in the first 15 generations) but >80-fold resistance after 24 generations, which has been confirmed in some populations of the whitefly (*Bemisia tabaci*) and the Colorado potato beetle (*Leptinotarsa decemlineata*).

Case Histories of Insecticide Resistance

India was one of the first country among third world countries to start a large scale use of synthetic pesticides for the control of insect pests of public health and agricultural importance. The modern era of vector control and plant protection in India started with the introduction of DDT in 1947 followed by HCH in 1949, organophosphates in 1953 and carbamates a little later. Despite the fact that these pesticides have brought immense benefits to the country, they also exhibited serious environmental consequences. It is interesting that DDT and HCH, which have been withdrawn from use in most of the advanced countries of the world, were still being used freely in India for public health.

San Jose Scale Resistant to Lime Sulphur

Melander's (1914) reported the first case of field-evolved insecticide resistance in San Jose scale to lime sulphur that certain populations of insects but not all of them were becoming less susceptible to sulphur-lime than they had been in the past though the chemical was documented to be very effective at killing scale insects previously. Surprisingly, it was found that 90% of the insects that he had sprayed in Clarkston had survived and even when he increased the amount of active ingredient by ten times, still 74% of them still survived. He was of the opinion that San Jose scale should have become acclimatized to a sulphur-lime environment. By consuming repeated small amounts of arsenic the body becomes immune to many times the normal lethal dose. Melander also predicted that entire populations would not become resistant as long as some non-resistant insects survived, because their non-resistant genes would be passed on to future generations. If only the resistant individuals survived to reproduce then resistant line might result after repeated sprayings. But always there are some scales missed by the spraying, and these, during the ten generations between sprayings, will produce a population in part, at least, non-resistant (Levin 2014).

Paradigm Shift to Biopesticides

The insecticide resistance in insects in major crops made the various stake holders to think alternative pest control strategies to safe guard the crops from ravaging insect pests. One such potential alternative is exploration of biopesticides of microbial origin for the management of insects. In India, the development of microbial entomopathogens as insecticides has involved notable successes and failures in the past two decades. India is a tropical and subtropical country with diverse pest and beneficial insect fauna, and with crop losses due to insect pests estimated at 17.5% (valued at US$17.3 billion). Several classes of customarily used insecticides are now restricted or prohibited due to their harmful effect on the environment, human health, and non-target organisms. Concurrently, the past two decades have witnessed a rise in the use of microbial biopesticides based on entomopathogenic organisms in India.

The global biopesticide market was estimated at approximately $3 billion, or 5% of total crop protection market, in 2013 and is expected to grow to more than $4.5 billion by 2023 (Olson, 2015). The value of biopesticides as a part of integrated pest management (IPM) programs has led to the recent increase in their use in India; biopesticides were recently estimated to comprise about 4.2% of the Indian pesticide market (Das 2014). However, market growth has been restricted by slow adoption, limited resources for large-scale production, and challenges associated with regulation and commercialization (Singh et al. 2016). Undoubtedly, microbial biopesticides play vital role in controlling the desirable pests and gaining interest among the population with advantages like non toxic mechanism, eco-friendly nature, efficacy and suitability in the Integrated Pest Management programmes unlike synthetic insecticides.

Biopesticides, an alternative to chemical pesticides, are typically derived from living organisms, microorganisms, and other natural sources pose less risk to people and the environment and hence gain worldwide attention as a new tool to kill insects. Biopesticides are being widely used to manage biotic stresses as a component of IPM under protected cultivation (Ramasamy and Ravishankar, 2018). On considering the international market of export commodities and health conscious of Western countries, the role of microbial biopesticides in pest management would address their importance among the growers and consumers which ultimately enhance the marketability of microbial biopesticides in India. Development and promotion of biopesticides usage in India need to be well addressed through promoting their manufacture at village level as an ancillary profession to agriculture. Registration process of biopesticides in India may be simplified without compromising quality and authenticity of the product. Policy decisions regarding production, development, promotion of biopesticides in India would definitely attract more scope in near future. A strict follow up of the policies pertaining to promotion and use of biopesticides will encourage inviting definite foreign exchange, besides producing healthy food commodities in India. Enormous scope for biopesticide market in India if the industry and extension functionaries convince farmers and pesticide manufacturing companies for better utilization of biopesticides in India as they are cheap, economical, viable, durable and effective.

Way Forward

The best way to overcome insecticide resistance is to reduce selection pressure and preserve the finite resources of new and useful compounds by adopting resistance management strategies in an integrated approach. Careful and systematic planning of insecticide application includes monitoring of resistance genes (or the associated enzymes or channels) in pest populations, as is now feasible for many of the mutated targets will help in partly in delaying the resistance. Resistance management is often necessary to shift to new compounds acting on novel targets that once again minimize selection pressure. This process of continually shifting approaches may ultimately be limited by a finite number of practical targets for pest control. Adopting the integrated pest management (IPM) approach usually helps with resistance management by retaining some susceptible populations along with resistant individuals.

The best way to delay onset of evolution of resistance in pests to minimize insecticide use and integration of chemical and non-chemical controls to seek safe, economical, and sustainable suppression of pest populations.The non-chemical approaches such include biological control by predators, parasitoids, and pathogens. Also valuable are cultural control through crop rotation, manipulation of planting dates to limit exposure to pests, and use of cultivars that tolerate pest damage and mechanical controls by exclusion using barriers and trapping. Tank mixing pesticides is the combination of two or more pesticides with different modes of action in order to improve individual pesticide application results and delay the onset of or mitigate existing pest resistance. Exploration and utilization of botanicals, biopesticides of biological origin to achieve target production as alternatives to synthetic chemical insecticides would serve the purpose. Educating, equipping and encouraging farmers to utilise biological pesticides to reduce the cost of protection, environmental pollution, without compromising yield and market price.

Conclusions

Although insecticides have been proved as one of the best management options, looking into growing demand for organic production and health consciousness of consumers across the world, adverse effect of environmental pollution and health hazards, biological based pest management strategies such as biological control, microbial control and have to be intensified and encouraged. Biopesticides, an alternative to chemical pesticides, are typically derived from living organisms, microorganisms, and other natural sources pose less risk to people and the environment and hence gain worldwide attention as a new tool to kill insects. Biopesticides are being widely used to manage biotic stresses as a component of IPM under protected cultivation. Development and promotion of biopesticides usage in India need to be well addressed through promoting their manufacture at village level as an ancillary profession to agriculture.

References

Guedes RNC (2017) Insecticide resistance, control failure likelihood and the First Law of Geography. Pest Management Science 73(3):479-484.

Ju D, Mota-Sanchez D, Fuentes-Contreras E, Zhang YL, Wang XQ and Yang XQ (2021) Insecticide resistance in the *Cydia pomonella* (L): Global status, mechanisms, and research directions. Pestic Biochem Physiol 178:104925.

OECD-FAO Agricultural Outlook (2012) Organisation of economic co-operation and development and Food and Agriculture Organization of the United Nations. https://www.oecd-ilibrary.org/agriculture-and-food/oecd-fao-agricultural-outlook-2012_agr_outlook-2012-en.

FAOSTAT, Food and Agriculture Organization of the United Nations. http://www.fao.org/faostat/en/#data/QC.

Alewu B, Nosiri C (2011)Pesticides and human health. In: Stoytcheva M, editor. Pesticides in the Modern World – Effects of Pesticides Exposure. InTech. p. 231–50. Available from:http://www.intechopen.com/books/pesticides-in-the-modern-world-effects-of-pesticides- exposure/pesticide-and-human-health.

Hoffman RS, Capel PD, Larson SJ (2000) Comparison of pesticides in eight U.S. urban streams. Environ Toxicol Chem 19:2249-58. doi:10.1002/etc.5620190915.

Nicolopoulou-Stamati P, Maipas S, Kotampasi C, Stamatis P Hens L (2016) Chemical pesticides and human health: the urgent need for a new concept in agriculture. Public Health Front 4:148.

Oppenoorth FJ, Welling W (1976) Biochemistry and physiology of resistance. In PesticBiochemPhysiol. Springer, Boston, MA. Boston, MA, 507-551pp.

Moyes CL, Vontas J, Martins AJ,(2017) Contemporary status of insecticide resistance in the major Aedes vectors of arboviruses infecting humans. PLOS Neglected Tropical Diseases11(7):e0005625.

Liu N (2015) Insecticide resistance in mosquitoes: impact, mechanisms, and research directions, Ann Rev Entomol 60(1):537-559.

Hemingway J, Hawkes NJ, McCarroll L, Ranson H(2004)"The molecular basis of insecticide resistance in mosquitoes". Insect Biochem Mol Biol 34(7):653-665.

Auteri M, La Russa F, Blanda V Torina A (2018) Insecticide resistance associated with kdr mutations in Aedes albopictus: an update on worldwide evidences. Biomed Res Int https://doi.org/10.1155/2018/3098575.

Fand BB, Nagrare VS, Gawande SP, Nagrale DT,Naikwadi, BV, Deshmukh V, Gokte-Narkhedkar N. and WaghmareVN (2019) Widespread infestation of pink bollworm, Pectinophoragossypiella (Saunders)(Lepidoptera: Gelechidae) on Bt cotton in Central India: a new threat and concerns for cotton production. Phytoparasitica 47(3):313-325.

Olson S (2015) An analysis of the biopesticide market now and where it is going. Outlooks on Pest Management 203-206. doi: 10.1564/v26_oct_04.

Das SK (2014) Recent development and future of botanical pesticides in India. Popular Kheti 2:93-99.

Singh HB, Keswani C, Bisen K, Sarma BK, Kumar P (2016) Development and application of agriculturally important microorganisms in India, in India 167-182pp. In: HB Singh, BK Sarma, C Keswani (eds), Agriculturally Important Microorganisms: Commercialization and Regulatory Requirements in Asia. doi:10.1007/978-981-10-2576-1-10

Ramasamy S, Ravishankar M (2018) Integrated pest management Brust, G.E., Perring, T.M., (Eds.), Sustainable Management of Arthropod Pests of Tomato, Elsevier 313-322. doi:: 10.1016/B978-0-12-802441-6.00015-2

Sahani SK, Pal S (2021). Insect Resistance to Neonicotinoids-Current Status, Mechanism and Management Strategies.

Kranthi, K. R. 2015. Pink Bollworm strikes Bt cotton. (Cotton Statistics and News published by the Cotton Association of India). 2015-16, No.35 1st (superscript) December 2015), 1-5pp.

Naveen NC, Chaubey R, Kumar D, Rebijith KB, Rajagopal R, Subrahmanyam B, Subramanian S (2017). Insecticide resistance status in the whitefly, Bemisia tabaci genetic groups Asia-I, Asia-II-1 and Asia-II-7 on the Indian subcontinent. Sci Rep 7(1):1-5

Kranthi KR, Jadhav DR, Kranthi S, Wanjari RR, Ali SS, Russell DA (2002)Insecticide resistance in five major insect pests of cotton in India. Crop Protect 21:449-460.

Kranthi KR, Russell D, Wanjari R, Kherde M, Munje S, Lavhe N, Armes N (2002) In-season changes in resistance to insecticides in Helicoverpa armigera (Lepidoptera: Noctuidae) in India. J Econ Entomol 95(1):134-42

Armes NJ, Jadhav DR, De Souza KRA (1996) survey of insecticide resistance in *Helicoverpa armigera* in the Indian subcontinent. Bull Entomol Res 86:499-514.

Melander L (1914) Can insects become resistant to sprays? J Econ Entomol 7:167-173.

Callaghan A (1991) Mechanisms and detection of insecticide resistance: a review. Science Progress 75:423-438.

Sekhar PR, Venkataiah M, Rao NV, Rao BR, Rao VSP (1996) Monitoring of insecticide resistance in *Helicoverpa armigera* (Hubner) from areas Receiving, heavy insecticidal applications in Andhra Pradesh (India). J Entomol Res 20:93-102.

Links

https://entomologytoday.org/2014/04/08/the-first-journal-article-on-insecticide-resistance-was-published-100-years-ago-this-month/downlaoaded dated 22 Dec 2021.

3

History and Development of Biological Control

Abstract

Chinese farmers were the first to apply biological control of agricultural pests when they used red ants to manage pests of fruit crops. Neem-based products were then used as fertiliser and as a barrier against pests of stored goods. Later, biopesticides surpassed macro-biocontrol agents in prominence and were added as one of the elements of integrated pest management. The interventions and directives of the United States Environmental Protection Agency and European Food Safety Authority were for way forward to the evolution of pesticides derived from naturally occurring organisms and plant materials. The development of biopesticides in agriculture across the world and India is reviewed in this chapter, along with the contributions of the Canada Department of Agriculture, Common Wealth Institute of Biological Control, Department of Biotechnology, New Delhi, Biotech Consortium India Ltd., New Delhi, etc. Since the Indian Council of Agricultural Research, New Delhi, launched its AICRP-Biocontrol programmes, a dramatic rise in the use of biopesticides has been observed in India.

Keywords: Microbial Biopesticide, History, USEPA, EFSA, ICAR initiatives

Introduction

The biocontrol use had almost completely disappeared due to the growth and success of the synthetic pesticide industry in the mid-1940s. The publication of Rachael Carson's 'Silent Spring' (Carson 1962) which condemned the use of agricultural pesticides and emphasized their harmful environmental effects on wildlife. Due to public outrage over this controversial book, there was a need for pesticide alternatives, which presents an opportunity for wider use of biological control (Barratt et al. 2010; Gay 2012). Many naturalists and environmentalists began looking for new insecticides with innovative chemical structures that would have less harmful effects on people, animals, and the environment (Barratt et al. 2018).

In 1901, Japanese biologist Shigetane Ishiwata discovered spores of the bacterium, *Bacillus thuringiensis* (*Bt*) from a sick silkworm. This bacteria is still the most often used biopesticide today (Chen 2014; Glare et al. 2000). Sporeine, the first *Bt* product to be sold commercially, debuted in 1938. The extensive usage of biopesticides started in the 1950s in the US. A low level of research and development was maintained in the second half of the 20th century as a result of the widespread use of synthetic chemical insecticides and World War II. The Pacific Yeast Product Company created the submerged fermentation industrial process in 1956, allowing for the large-scale manufacturing of *Bt* (Glare et al. 2000). In 1973, *Heliothis* NPV was granted exemption from tolerance and the first viral insecticide, Elcar received a label in 1975. In 1977, *B. thuringiensis* var. israelensis, which is poisonous to flies, was reported in 1977, while the strain tenebrionis, which is poisonous to beetles, was found in 1983. Following the demonstration by environmentalists and ecologists that widespread and repeated application of these synthetic chemicals could be ecologically detrimental, biological pest management was nevertheless expedited (Cook and Baker 1983).

Earlier, biocontrol agents like some predatory insects (red ant) and birds were engaged for insect pest management (Brahmachari 2004). Later few botanicals including various parts of neem tree (*Azadirachta indica* A. juss) and its extracts were tried as fertilizers and also to protect stored cereals from post-harvest losses (Isman 1997; Schmutterer 1985).

The concept of Integrated Pest Management (IPM) had come to the field of Agriculture during 1960s, in which judicious use of various methods of control was emphasized to overcome the ill-effects of chemical pesticides (Smith and Bosch 1967). Later on, based on the recommendation of US National Academy of Sciences, biological control with natural enemies and microbial biopesticides was included one of the components in IPM (Peshin et al. 2009). Control failure of few polyphagous cotton feeders including American boll worm, *Helicoverpa armigera* and generalized defoliator, *Spodoptera litura* with chemical pesticides alerted to switch over biological control, a safe, cost-effective, and eco-friendly method (Kranthi et al. 2002).

By the mid 1920s, entire British Empire was active in biological control work including Australia, New Zealand, Fiji, Canada, Bermuda and South Africa. In 1927, the Imperial Bureau of Entomology (IBE) created facilities for conducting biological control work in Farnham House Lab, England. It was under the control of W. R. Thompson in 1928 who initially concentrated on natural enemies of insect pests and broadened to work on biological control of weed in 1929. In 1929, Canada Department of Agriculture (CDA) constructed a biological control lab at Belleville, Ontario. In 1940, this lab was moved to Ottawa, Canada where it became as Imperial Parasite Service. In 1947, it became independent and designated as Commonwealth Bureau of Biological control (CBBC). In 1951, it was renamed as Commonwealth Institute of Biological Control (CIBC). In 1961, the CIBC headquarters were transferred to Trinidad, West Indies. The CIBC identified two sub-stations in south East Asia, one at Bengaluru in India and another at Rawalpindi in Pakistan to undertake biological control research. In 1957, the India station of CIBC was established to initiate organized and systematic research in biological control.

Aristotle was the first to mention in his book "Historia Animalia" that honey bees suffered a disease which was later identified as foul brood disease. One chapter on diseases of Insects was included in the book entitled 'An Introduction of Entomology'. Disease can be defined as a departure of the insect from a state of health and was first noticed among domesticated insects. In Europe, Aristotle was the first to mention that bees suffered disease and in 1835, Agostino Bassi showed that animal disease could be caused by a microorganism, when he found that the fungus *Beauveria bassiana* causes the muscardine disease of silkworms. Early observations were largely concentrated on two domesticated insects, the honey bee and silkworm. Gradually these studies were extended to pest species too, and the concept of utilizing disease to control these insects was born.

In 1879, the Russian, Metchnikoff, conducted the first significant experiments in the destruction of injurious insects with micro organisms by infecting larvae of the beetle *Anisoplia austriaca* with the fungus *Metarhizium anisopliae*. The first commercial product, Sporeine, containing *Bacillus thuringiensis* was produced before 1938. After the second world war, several commercial firms in the USA, began to produce this bacterium. Today we can point to such achievements as the protection of over 50% of the cole crops from the cabbage looper, *Trichoplusia ni* in Southern California by *B. thuringiensis* in 1965 and 1966. The importance of efficient, eco-friendly methods for pest and disease control gained momentum. Steady growth of biological control has been reported in various eras including Ancient origins, North American Beginnings, California origins and 20th Century developments with several explorations and examples for successful management of many key pests of crops.

20th and 21st Century Developments

Many chemical pesticides were withdrawn from the market in the second half of the 20th century as a result of inappropriate application techniques used during World War II, such as aerial application (Howard 1935), which had numerous negative consequences, including acute or chronic toxicity (Hunt and Bischoff 1960), as well as other unfavorable effects like increased resistance in the target species (Mouches et al. 1986), the replacement of target species with more dangerous resistant species (Regnault-Roger et al. 1986), and contamination of different environmental compartments (Ellgehausen et al. 1980; Leduc et al. 1987). Authorities from the US Environmental Protection Agency (USEPA) and the European Food Safety Authority (EFSA) revised the pesticide laws when these issues arose in order to protect human and animal health as well as the environment from the risks associated with pesticides. They proposed the ideas of the ideal pesticide, which include i) a high selectivity to target species but a minimal toxicity to non-target organisms, ii) a high effectiveness at a low application rate, and iii) a low environmental persistence. Thus a new concept on 'Biopesticide' had evolved to fight with pests effectively but have minimal impacts on humans, animals, and the environment. Active biopesticide research has expanded in the latter decade of the 20th century along with a notable increase in publications (Shukla et al. 2019) (Fig. 1).

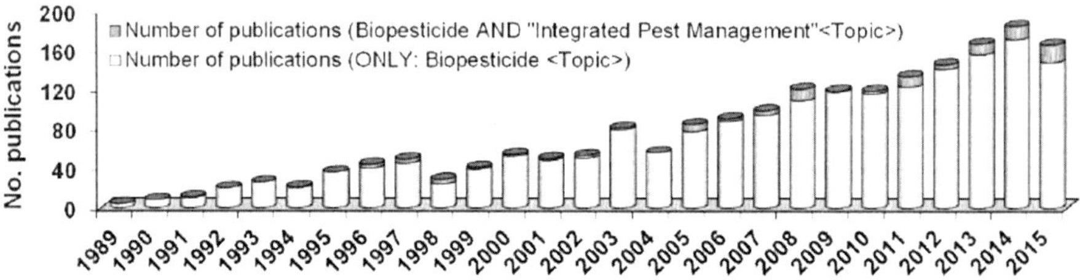

Fig. 1. Publications on biopesticide research from 1989 to 2015 (Web of Science 2015).

Around 1400 biopesticide products were sold globally at the start of the twenty-first century, making up around 2.5% of the entire pesticide business (Marrone 2007; Chandler et al. 2012). From 2012 to 2017, the demand for biopesticides was predicted to increase at a faster compound annual growth rate (GAGR) of 16.1% (compared to 3%) than that of synthetic pesticides, resulting in an estimated $ 5.2 billion global market in 2017 (Lehr 2010). Europe, Oceania, and Latin America accounted for 20, 20, and 10% of worldwide biopesticide consumption, respectively, whereas North America consumed roughly 40% of them (Leng et al. 2011). Numerous legislative initiatives for the sustainable use of pesticides were implemented, and they all emphasized how crucial it is to use less chemical pesticides overall to avoid potential environmental damage.

When the All India Coordinated Research Project on Biological Control of Crop Pests and Weeds (AICRPBC) was established in 1977, the biological control effort received a further boost. In 1993, the AICRPBC was elevated even further to the position of Project Directorate of Biological Control (PDBC). The goal of the PDBC was to do basic and applied research on the biological control of crop pests and weeds across the nation. With 16 sites dispersed throughout the nation, PDBC serves as the central agency in India. The name of PDBC has been modified twice. PDBC was upgraded to become the National Bureau of Agriculturally Important Insects (NBAII) during the XIth 5-year plan (2009), and in the XIIth 5-year plan, it was renamed as National Bureau of Agricultural Insect Resources (NBAIR).

In 1989, the National Bio-control Network Program was also introduced by the Department of Biotechnology (DBT) in New Delhi. Ten R&D projects were launched at the beginning of the programme with a duration of five years (1989-1994). Over 200 projects were carried out in numerous national institutes and state agricultural universities (SAUs) once the programme was later expanded (Wahab 2004). IPM was included in the government of India's National Policy Statement in 1985, but the Department of Agriculture and Cooperation (DAC), Ministry of Agriculture, took a significant step by announcing a programme on "Strengthening and Modernization of Pest Management Approach in India in 1991-1992," along with the establishment and strengthening of biocontrol research at the regional level. Department of Biotechnology is one of the top financing organisations in India with programmes for biocontrol research (Singh et al. 2002; Sharma et al. 2003; Mishra et al. 2020). Currently, in addition to DBT, other funding organisations like the Department of Science and Technology (DST), New Delhi, and the Indian Council of Agricultural Research (ICAR) have also introduced a number of programmes with a major focus on prudent pesticide use in agriculture and supporting research on biopesticides. These government organisations are, however, also highly concerned with accumulating toxicological information regarding the use of biopesticides in diverse agro-climatic zones.

The first commercial biocontrol agent manufacturer in India was Bio-Control Research Laboratories (BCRL), a branch of Pest Control India (PCI) Limited working under a contract with the Plant Protection Research Institute (PPRI) (Manjunath 1992). Currently, *Trichoderma viride, Trichoderma harzianum*, and *Beauveria bassiana* are antagonistic bacteria and fungi that are manufactured and sold by the BCRL. Later, IPM was backed by the National Policy on Agriculture (2000) and the National Policy for Farmers (2007). In India, a total of 970 biopesticide formulations have been registered in Central Insecticide Board and Registration Committee (CIB&RC) as on 1.1.2020 in which *Trichoderma* and *Pseudomonas* based formulations dominate (http://164.100.161.213/divisions/cib-rc/bio-pesticide-registrant). A compendium consisting of 31 microbial formulations which are in different stages of registration and commercialization were reported to possess 3 to 25 months shelf-life at 25°C to 35°C (Saxena et al. 2021).

Conclusions

The history and expansion of biopesticide in agriculture around the world clearly showed that a phenomenal growth was reported during the second half of the 20th century and the beginning of the 21st century. The United States Environmental Protection Agency and the European Food Safety Authority's interventions, which involved revising the laws, greatly aided in the registration of plant protection products that contained less harmful ingredients to replace conventional pesticides or serve as a starting point for the synthesis of novel chemistries. The development of biopesticides was thought to be based on extraction from natural sources, chemical synthesis, and computational chemistry. Additionally, biopesticides are not harmless and are subject to the same laws and regulations as chemical pesticides (Villaverde et al. 2016). More than 900 biopesticide formulations have been certified through the Central Insecticide Board and Registration Committee, New Delhi. Over 31 bacterial and fungal-based biopesticide formulations are in pipeline to pass various stages of commercialization in India.

References

Barratt BIP, Howarth FG, Withers TM, Kean J, Ridley GS (2010) Progress in risk assessment for classical biological control Biol Control 52:245-254.

Barratt BIP, Moran VC, Bigler F, van Lenteren JC (2018) The status of biological control and recommendations for improving uptake for the future. BioControl 63:155-167.

Brahmachari G (2004) Neem-an omnipotent plant: a retrospection. Chem Bio Chem 5(4):408-421.

Carson R (1962) Silent spring, Hamish Hamilton, London.

Chen R (2014) Biopesticides: A Formulator's Perspective. http://news.agropages.com/News/NewsDetail-14156.

Cook R, Baker K (1983) The Nature and Practice of Biological Control of Plant Pathogens. St Paul, MN: Am. Phytopathol. Soc.

Ellgehausen H, Guth JA, Esser HO (1980) Factors determining the bioaccumulation potential of pesticides in the individual compartments of aquatic food-chains. Ecotox Environ Safe 4(2):134-157.

Gay H (2012) Before and after silent spring: from chemical pesticides to biological control and integrated pest management-Britain, 1945-1980. Ambix 59(2):88-108.

Glare TR, O'Callaghan M (2000) *Bacillus thuringiensis*; Biology, Ecology and Safety, Wiley

Howard LO (1935) *La menace des insects*, Flammarion, Paris, France.

Hunt EG, Bischoff AI (1960) Inimical effects on wildlife of periodic DDD applications to Clear Lake'. Calif Fish Game 46:91-106.

Isman MB (1997) Neem and other botanical insecticides: barriers to commercialization. Phytoparasitica 25(4):339.

Kranthi KR, Russell D, Wanjari R, Kherdc M, Munje S, Lavhe N, Armes N (2002) In-season changes in resistance to insecticides in *Helicoverpa armigera* (Lepidoptera: Noctuidae) in India. J Econ Entomol 95(1):134-142.

Leduc R, Unny TE, McBean EA (1987). Stochastic modeling of the insecticide fenitrothion in water and semiment compartments of stagnant pond. Water Resour Res 23(7):1105-1112.

Manjunath TM (1992) Biological control of insect pests and weeds in India: notable successes. In: Hirose Y (ed) Biological control in South East Asia Kyushu Univ Press/IOBC/SEARS, Tokyo, 11-21pp.

Mishra J, Dutta V, Arora NK (2020) Biopesticides in India: technology and sustainability linkages. 3 Biotech 10:210. doi.org/10.1007/s13205-020-02192-7.

Mouches C, Paasteur N, Berge JB, Hyrien O, Raymond M, Desaintvincent BR, Desilvestri M, Georghiou GP (1986) Amplification of an esterase gene is responsible for insecticide resistance in a California *Culex* mosquito. Science 233(4765):778-780.

Peshin R, Bandral RS, Zhang W, Wilson L, Dhawan AK (2009) Integrated pest management: a global overview of history, programs and adoption. In: Peshin R, Dhawan AK (eds) Integrated pest management: innovation-development process. Springer, Dordrecht, 1-49pp.

Regnault-Roger C, Philogene BJR, Vincent C (2005) Biopesticides of plant origin, intercept Ltd, Wimborne, UK.

Saxena AK, Chakdar H, Kumar M, Rajawat MVS, Dubey SC, Sharma TR (2021) ICAR Technologies: Biopesticides for Eco-friendly pest management. Indian Council of Agricultural Research, New Delhi, India 1-31pp.

Schmutterer H (1985) The neem tree *Azadirachtia indica* A. Juss and other meliaceous plants: sources of unique natural products for integrated pest management, medicine, industry and other purposes. VCH Weinheim, Germany 696p.

Sharma M, Charak K, Ramanaiah T (2003) Agricultural biotechnology research in India: status and policies. Curr Sci 84(3):297-302.

Shukla N, Singh EANA, Kabadwa BC, Sharma R, Kumar J (2019) Present status and future prospects of bio-agents in agriculture. Int J Curr Mirobiol App Sci 8(4):2138-2153.

Singh RS, Singh PP, Bedi JS (2002) Final report of DBT scheme on biocontrol of seed and soil-borne diseases of vegetables. Punjab Agricultural University, Ludhiana, 90p.

Smith RF, van den Bosch R (1967) Integrated control. In: Kilgore WW, Doutt RL (eds) Pest control: biological, physical and selected chemical methods. Academic Press, Inc., New York, 295-340pp.

Villaverde JJ, Sandin-Espana P, Sevilla-Moran B, Lopez-Goti C, Alonso-Prados JL (2016) Biopesticides from natural products: Current development, legislative framework, and future trends. Bioresources 11(2):5618-5640.

Wahab S (2004) The Department of Biotechnology initiates towards the development and use of biopesticides in India. In: Kaushik E (ed) Biopesticides for sustainable agriculture: prospects and constraints. The Energy and Resources Institute, New Delhi, 73-90pp.

Links

http://164.100.161.213/divisions/cib-rc/bio-pesticide-registrant

of Plant Protection Quarantine & Storage in Lucknow, are major government agencies which are mass producing biopesticides in Uttar Pradesh. Many ICAR sponsored Krishi Vigyan Kendras (KVK), State Government sponsored state biocontrol labs and National Agricultural Co-operative Marketing Federation of India (NAFED) are in the full time job in production of biopesticides.

In India, public sectors contribute 70% of the biopesticides production. Major companies are Biotech International Ltd., New Delhi, International Panaacea Ltd, New Delhi, Ajay Biotech (India) Ltd., Pune, Bharat Biocon Pvt. Ltd., Chhattisgarh, Microplex Biotech and Agrochem Pvt., Mumbai, Excel Crop Care Ltd., Mumbai, Govinda Agro Tech Ltd., Nagpur, Jai Biotech Industries, Satpur, Nasik, Ganesh Biocontrol System, Rajkot, Gujarat Chemicals and Fertilizers Trading Company, Baroda, Gujarat Eco Microbial Technologies Pvt. Ltd., Vadodara, Chaitra Agri-Organics, Mysore, Deep Farm Inputs (P) Ltd., Thiruvanandapuram, Kerala, Kan Biosys Pvt. Ltd., Pune, Indore Biotech Inputs and Research Pvt. Ltd., Indore, Romvijay Biotech Pvt. Ltd., Pondichery, Devi Biotech (P) Ltd., Madurai, Tamil Nadu, T. Stanes and Company Ltd., Coimbatore, Tamil Nadu, Harit Bio Control Lab., Yavatmal and Hindustan Bioenergy Ltd., Lucknow. Few Indian companies which work in biopesticde production in collaboration with foreign companies are Lupin Agro-chemicals, Mumbai; Sugar and distillery companies such as KCP Sugar and Industries Corporation Ltd., Andhra Pradesh, Rajshree Sugars and Chemicals Ltd., Tamil Nadu; New Swadeshi Sugar Mills, Bihar, and Bannari Amman Sugars Ltd., Tamil Nadu.

Biopesticide Registrant in India

As on 1.1.2021, a total of 970 biopesticides registered in India by CIB&RC under the 1968 Insecticide Act which include microbial biopesticides of *Bacillus thuringiensis* var. kurstaki (42), var. israelensis (22), var. sphaericus (05), var. galleriae (01), *Pseudomonas fluorescence* (196), *Bacillus subtilis* (04), *Trichoderma viride* (289), *T. harzianum* (51), *Ampyliomyces quisqualis* (02), *Beauveria bassiana* (106), *Metarhizium anisopliae* (30), *Verticillium lecani* (93), *Verticillium chlamydosporium* (03), *Helicoverpa armigera* NPV (30) and *Spodoptera litura* NPV (03) (Kumar et al. 2018; Keswani et al. 2019; http://ppqs.gov.in/divisions/cib-rc/biopesticide-registrant) (Table. 1).

Table 1: List of microbial biopesticide formulations registered in Indian

S.No.	Name of Company	Organism	Trade name	Target biotic stress
1.	M/s Rallis India Ltd., Bengaluru	*Bacillus thuringiensis*	Bobit IIWP	Lepidoptera insects
2.	M/s Sandoz (I) Ltd., India	*Bacillus thuringiensis* var. kurstaki	Deflin WG	Lepidoptera insects
3.	M/s Aventis Crop science Ltd., Bengaluru	*Bacillus thuringiensis* var. israelensis (H-14)	Vectobac 12 AS	Diptera insects
4.	M/s T. Stanes and Companey Ltd., Cotore, India	*Beauveria bassiana*	Biopowder WP	Sucking insects, bollworms
5.	M/s Biotech Industries Ltd., New Delhi	NPV	Biovirus-H	*Helicoverpa armigera*
6.	M/s Pest control India Ltd., Bengaluru	NPV	Spodocide 0.50% AS	*Spodoptera litura*
		NPV	Helicide 0.50% AS	*Helicoverpa armigera*
7.	M/s Anshul Agrochemicals, Bengaluru	*Beauveria bassiana*	Green Heal	Sucking insects, borers, bollworms
		Beauveria bassiana	Beveroz-L	Sucking insects, borers, bollworms
		Beauveria bassiana	Almax	Sucking insects, borers, bollworms
		Pseudomonas florescens	Pseudomax	Soil and seed borne diseases
		Trichoderma viridi	Trichomax	Soil borne diseases, plant parasitic nematodes
8.	M/s Deepa Farm inputs Private Limited	*Pseudomonas florescens*	Bio-Plus Pseudo	Soil and seed borne diseases
		Trichoderma viridi	Bio-Plus Tricho	Soil borne diseases, plant parasitic nematodes
		Verticilium lecanii	Bio-Plus Verticillium	Sucking pests, plant parasitic nematodes
		Metarhizium anisopliae	Bio-Plus Metarhizium	Beetles, soil arthropods
9.	M/s Kan biosys Pvt. Ltd., Pune	*Trichoderma harzianum*	Nemastin 1% WP	Root knot nematode
		Trichoderma viridi	Combat 1% WP	Root rot, damping off, wilt
		Beauveria bassiana	Brigade-B 1.15% WP	Rice leaffolder
		Pseudomonas florescens	Sudo 0.5% WP	Late leaf spot of groundnut

S.No.	Name of Company	Organism	Trade name	Target biotic stress
10.	M/s Peak Chemical Industries Ltd., West Bengal	*Paecilomyces fumosoroseus*	Bardan	Spider mite, parasitic mites
		Metarhizium anisopliae	Moti	Termite
		Beauveria bassiana	Badsha	Sucking pests
		Verticilium lecanii	Victor	Parasitic nematodes, whitefly, thrips, aphids
11.	M/s Uttam Chemicals Industries, Rajasthan	*Trichoderma harzianum*	Trichoderma harzianum 1% WP	Parasitic nematodes
		Trichoderma viridi	Trichoderma viridi 1.5% WP	Nematicide, crop diseases
12.	M/s Ambic Organic, Surat	*Paecilomyces fumosoroseus*	Almite	Mites, DBM, sucking pests
13.	M/s Criyagen Agri & Biotech Pvt. Ltd., Bengaluru	*Trichoderma viridi*	Trichoderma	Soil borne fungus
14.	M/s Biotech International Ltd., New Delhi	*Bacillus thuringiensis* var. kurstaki	BIOLEP WP	*Helicoverpa, Spodoptera*, DBM, borers, hairy caterpillars, cut worms, army worms, leaf rollers & miners, skeletonizers & Defoliators
		Beauveria bassiana	BIORIN WP/AS	*Helicoverpa, Spodoptera*, DBM, leaf borer, hairy caterpillars, mites, spidermites, whiteflies, aphids, scale insects, locusts & colorado beetles
		Verticillium lecanii	BIOLINE WP/AS	Whitefly, green leaf hopper, thrips, mealy bug, brown plant hopper, leaf miner, aphids, mites, jassids
		Metarhizium anisopliae	BIOMET WP/AS	White grubs, termite, cut worm, caterpillars, semiloopers, sucking pests, mealybugs & aphids

S.No.	Name of Company	Organism	Trade name	Target biotic stress
		NPV	BIOVIRUS-H AS	*Helicoverpa armigera*
		NPV	BIOVIRUS-S AS	*Spodoptera litura*
		Pseudomonas fluorescens	BIOMONAS WP/AS	Bacterial wilt, black rot, bacterial spot, sheath blight; blast, anthracnose, powdery & downy Mildew, panama wilt, panama wilt, sigatoka, bacterial leaf spot
		Trichoderma viride	BIODERMA WP/AS	Root rot, stem rot, damping off, fusarium wilt and verticillium wilt, all types of leaf spot, leaf & blight
		Trichoderma harzianum	BIODERMA-H WP/AS	Root rot, stem rot, damping off, fusarium wilt and verticillium wilt, all types of leaf spot, leaf & blight
		Bacillus subtilis	BIOSUBTILIN WP/AS	Fusarium wilt, Macrophomina, damping off, Pythium, Rhizoctonia Black scarf of Potato, root rot, leaf spot, powdery and downey mildew, bacterial spot & bacterial leaf blight
		Ampelomyces quisqualis	ARMOUR WP/AS	Powdery Mildew in pulses, vegetables, fruits & ornamental crops
		Paecilomyces lilacinus	BIONEMAT WP/AS	Root-knot nematodes, reniform nematodes, cyst nematode, golden cyst nematodes, citrus nematodes, lesion nematode
		Bacillus firmus	NEMATO CURE WP/AS	Root-knot nematodes, reniform nematodes, cyst nematode, golden cyst nematodes, citrus nematodes, lesion nematode
		Hirsutella thompsonii	NO MITE WP/AS	Various types of crop mite – Scoulet, purple, red spider mite

S.No.	Name of Company	Organism	Trade name	Target biotic stress
15.	M/s International Panaacea Ltd, New Delhi	*Trichoderma viride*	Sanjeevni 1.0% WP	*Fusarium*, Charcoal rot, Black scurf, Karnal bunt, Silver leaf of plum & peach, *Rhizoctonia, Pythium, Schlerolims, Verticillium, Alternaria*
		Trichoderma harzianum	Bioharz 2% AS	*Fusarium*, Charcoal rot, Black scurf, Karnal bunt, Silver leaf of plum & peach, *Rhizoctonia, Pythium, Sclerolims, Verticillium, Alternaria*
		Pseudomonas flourescens	Rakshak 1% WP	Soil and seed borne diseases
		Trichoderma viride	Bokashi Bran	Soil and seed borne diseases
		Trichoderma viride	Seed2plant	Soil and seed borne diseases
		Trichoderma viride	Tricho-PEP H 1% WP	Diseases
			Trichoz-P 1.5% WP	root rot, wilt, stem rot
		Paecilomyces lilaecinus	Nematofree 1% WP	*Helicoverpa, Spodoptera*, borers, hairy caterpillars. pest of vegetables and fruits, whitefly, aphids, DBM, scale insects, locust, Colorado potato beetles
		Pseudomonas fluorescens	Bactvipe 2% AS	Root rot, stem rot, collar rot, wilt, blights, leaf spots, anthracnose, Alternaria and downy & powdery mildews
		Ampelomyces quisqualis	MilGO 2% AS	Powdery mildew, *Alternaria solani, Colletotrichum, Coccodes, Cladosporium, cucumerinum*
		Beauveria bassiana	DAMAN 1% WP	*Helicoverpa, Spodoptera*, borers, hairy caterpillars, pest of vegetables and fruits, whitefly, aphids, DBM, scale insects, locust, Colorado potato beetles
		Bacillus subtilis	MILDOWN 2% AS	*Pythium, Alternaria, Xanthomonas, Rhizoctonia, Botrytis, Scelerotiana, Phytophora*

S.No.	Name of Company	Organism	Trade name	Target biotic stress
16.	M/s Ajay Biotech (India) Ltd, Pune	*Bacillus thuringiensis* var. Kurstaki	Bio Dart	Insecticide
		Metarhizium anisopliae	Meta Guard WP/AS	Termites, white grubs, soil insects
		Beauveria bassiana, Metarhizium anisopliae, Verticillium lecanii	AJAY VBM	Insect pests
		Paecilomyces spp.	Nemadart	Parasitic nematodes
		Consortia of microbes	TERMiNIL	Termites
		Trichoderma viride	Trident 1.5% WP	Soil and seed borne diseases
		Trichoderma harzianum	Trichoshield	Soil borne fungus
		Trichoderma viride	ROM Trichoderma	Rhizome rot of turmeric and cardamom, fusarium wilt of banana, wilt disease of pepper, beetelvine, chillies, tomato and vegetables, *Phytophthora, Pythium, Sclerotium*
17.	M/s Bharat Biocon Pvt. Ltd., Chhattisgarh	*Pseudomonas fluorescens*	Pseudocon 0.5% WP	Crop diseases
		Trichoderma harzianum	TRICHOCON 1% WP	Crop diseases
		Beauveria bassiana	BASICON 1.15%	Lepidoptera insects
		Verticillium lecanii	VERTICON 1.15%	Plant parasitic nematodes
		Metarhizium anisopliae	METACON 1.15%	Beetle pests, locust
18.	M/s Jai Biotech Industries, Satpur, Nasik	*Verticillium lecanii*	Vertimust 1.1% WP	Sucking pests
		Beauveria bassiana	Beauvera 1.15% WP	Lepidoptera pests
		Trichoderma viride	Jaimold 1% WP	*Sclerocium, Rhizoctonium, Pythium, Fusarium*

S.No.	Name of Company	Organism	Trade name	Target biotic stress
19.	M/s Ganesh Biocontrol System, Rajkot	NPV	Biokills	*Spodoptera litura*
		Pseudomonas fluourescens	Monas	*Rhizoctinia* , *Sclerotini*, blights & Alternaria, *Ascochyta, Cercospora, Macrophomina, Xanthomonas, Erwinia*
		Beauveria bassiana	BASS 1.5% WP	*Helicoverpa, Spodoptera,* DBM, borers, hairy caterpillar of vegetables & fruit plants, mites & spidermites of vegetables & ornamentals , whiteflies on cotton & vegetables. aphids & scale insects, locust, potato and Coffee pod - borer
20.	M/s Gujarat Eco Microbial Technologies Pvt. Ltd., Vadodara	*Trichoderma viride*	NR-III	Soil & air borne pathogens, sucking pest
		Trichoderma viride	TRIOJEET	Damping off, wilt, root rot
21.	M/s Indore Biotech Inputs and Research Pvt. Ltd., Indore	NPV	Helicop AS	*Helicoverpa armigera*
		Metarhizium anisopliae	BIO-MAGIC	Grub, termite, fruit flies, hopper, wire worms, vegetable worms, aphids, jassids
		Beauveria bassiana	BIO-WONDER 1.15% WP	Hairy insects, aphids, white flies, mealy bugs, grasshoppers, thrips, stem borer, termites, beetles, caterpillars
		Verticillium lecanii	Vercitile 1.15% WP	Aphids, whiteflies, thrips, mealy bugs, scale insects, leaf hopper, mango hopper
		Bacillus thuringiensis var. kurstaki	Cezar 0.5% WP	Castor semilooper, soybean and gram pod borer, *Spodoptera litura*, Bihar hairy caterpillar, sphinx moth
		Paecilomyces lilacinus	BioAce 1% WP	Plant parasitic nematodes
		Trichoderma viride	Biohit WP	*Pythium, Fusarium oxyforum, Rhizoctonia solani, Alternaria, Sclerotinia rolfsia*
		Pseudomonas fluorescens	Biomonarch	*Macrophomina, Fusarium, Rhizoctonia, Sclerotium, Pythium*

S.No.	Name of Company	Organism	Trade name	Target biotic stress
22.	M/s Romvijay Biotech Pvt. Ltd., Pondichery	*Hirsutella thompsoni, Verticillium lecanii*	Biomite	All species of mites
		Beauveria, Bacillus subtilis	ROM Grub kill	Beetles, borers
		Metarhizium anisopliae	ROM Meta kill	Beetles, borers
		Beauveria bassiana	ROM Beevicide	Beetles, borers
		Paecilomyces lilacinus	ROM Pelicide	Plan parasitic nematodes
		Verticillium lecanii	VERELAC	Sucking pests
		Trichoderma viride	ROM Trichoderma	Wilt, rot, plant parasitic nematode
		Paecilomyces lilacinus	ROM TRY PAE MIX	Wilt, plant parasitic nematode
		Ampelomyces quisqualis	ROM No-Mildew	Mildew diseases
		Pseudomonas fluorescens	ROM Pseudomonas	Foliar and soil borne diseases
23.	M/s Devi Biotech (P) Ltd., Madurai, Tamil Nadu	*Trichoderma viride*	Boom Derma 1.5% WP	Damping off, wilt, collar rot, root rot, leaf blights spots
		Paecilomyces lilacinus	Boom Nemo 1% WP	Root knot nematodes, cyst nematodes, reniform nematode, burrowing nematode, citrus nematode, golden cyst nematode and lesion nematodes
		Pseudomonas fluorescens	Boom Monas 1% WP	Root rot, wilt, blast, sheath blight, damping off, leaf spot and rhizome rot
		Verticillium lecanii	Boom Vert 1.5% WP	Aphids, thrips, mealy bugs, white flies, jassids, Hoppers, scales and all types of mites.
		Beauveria bassiana	Boom Bass 1.15% WP	Root grubs, boll worms, *Spodoptera*, coffee berry borers, pod borers, hoppers and weevils.

S.No.	Name of Company	Organism	Trade name	Target biotic stress
24.	M/s T. Stanes and Company Ltd., Coimbatore, Tamil Nadu	*Verticillium lecanii*	BIO CATCH 1.15% WP	whiteflies, jassids, aphids, thrips, mealybugs
		Beauveria bassiana	Bio-Power 1.15% WP	Borers, cutworms, root grubs, leafhoppers. whitefly, aphids, thrips, mealybug
		Metarhizium anisopliae	Bio Magic 1.15% WP	Leaf hoppers, grasshoppers, root grubs, corn root worms, bugs, beetles, palm weevils, borers, cutworms, termites
		Paecilomyces lilacinus	Bio Nematon 1.15% WP	Root knot nematodes, burrowing nematodes, cyst nematodes, lesion nematodes
		Trichoderma harzianum	Bio Wrap 1% WP	Root-knot nematode, wilt disease of tomato, okra crops
		Entomopathogenic nematode	Crown	Root grub
		Streptomyces spp.	Stanomyte 1.5% LF	Mites
25.	M/s Harit Bio Control Lab., Yavatmal	NPV	Helistop	*Helicoverpa armigera*
		Trichoderma viride	Haritderma 1% WP	Damping, wilting, root spots, leaf spots and blights
		Verticillium lecanii	Versatile 1% WP	Sucking pests
		Beauveria bassiana	Wow 1.5% WP	Leaffolders
26.	M/s Bannari Amman Sugars Ltd., Tamil Nadu	*Bacillus licheniformis*	LEAF GUARD	*Actinopelte* Leafspot, Alternaria, Leafspot, leaf blight, Anthracnose, leaf blotch, Drechslera ink spot, Bipolaris Leaf spot, *Rhizoctonia* blight
		Trichoderma viride *T. harzianum*	ROOT CARE	Soil borne diseases
		Pseudomonas fluorescens	Pseudo Care	Crop diseases
		Bacillus subtilis	LEAF CARE	Fungal diseases

Table 2: List of biopesticides in pipe-line for registration and licensing in India

S.No.	Entomopathogenic fungi	Formulation	Shelf life	Trade name	Target pests	Dose
1.	*Beauveria bassiana* (Bals.-Criv.) Vuill. (1912) (Bb-5a)	Oil formulation (1×10^8 cfu/ml)	12 months at 25-35°C	Shatpada Aphid Kill	Chilli and brinjal aphids, *Aphis gossypii* Glover, 1877, cabbage aphid, *Brevicoryne brassicae* (Linnaeus, 1758); cowpea aphid, *A. craccivora* C. L .Koch, 1854	5 ml/lit. of water at 15 days interval
2.	*Isaria fumosorosea* Wize (1904) (Pfu5)	Talc (1×10^8 cfu/g); oil formulation (1×10^8 cfu/ml)	12 months at 25-35°C	Shatpada Rugose Whitefly Kill	Coconut and oil palm Rugose spiraling whitefly, *Aleurodicus rugioperculatus* Martin	2-3 foliar spray at 5 ml or 5 g/lit. of water at 15 days interval
3.	*Lecanicillium lecanii* R. Zare & W. Gams, 2001 (VI-8)	Oil formulation (1×10^8 cfu/ml)	12 months at 25-35°C	Shatpada Sucking pest Hit	Chilli aphids, *A. gossypii*, cowpea aphid, *A. craccivora*	Three foliar sprays at 5 ml/lit. of water at 15 days interval
4.	*Metarhizium anisopliae* (Metchnikoff) Sorokin (1883) Ma 4	Talc (1×10^8 cfu/ ml)	12 months at 25-35°C	Shatpada Grubicide	Sugarcane white grub, *Holotrichia* spp.	Soil application twice in a year during June/July, July/ August at 30 days interval at 2.5 kg mixed with 250 kg FYM/ha
5.	*M. anisopliae* Ma 35	Talc (1×10^8 cfu/g); oil formulation (1×10^8 cfu/ml)	12 months at 25-35°C	Shatpada Larvicide	Maize fall armyworm, *Spodoptera frugiperda* (J. E. Smith)	Three foliar sprays at 5 ml or 5 g/lit. of water at 20,30,40 days after sowing
6.	*Trichoderma reesei* Simmons, 1977 CST-T-3	Wettable powder (1×10^8 cfu/g)	12 months at 25-35°C	ICAR-Fusicont	Fusarium wilt of banana, *Fusarium oxysporum* f. sp. *cubense* E. F. Sm., W. C. Snyder & H. N. Hansen (1940) Tropical race 4 and race 1	Soil drenching at 1 lit/, four times on 3rd, 5th, 9th, 12th month after planting

S.No.	Entomopathogenic fungi	Formulation	Shelf life	Trade name	Target pests	Dose
7.	*T. harzianum* Rifai, (1969) + *Bacillus amyloliquefaciens* Priest *et al.*, 1987	Talc (1×10^7 cfu/g)	12 months at 25-35°C	Bio-Pulse	Wilt of chickpea, lentil, pea, pigeonpea; damping off/ seedling mortality in papaya; Target fungi (*Rhizoctonia, Sclerotium, Sclerotinia, Fusarium, Pythium, Ralstonia, Macrophomina, Bipolaris* , *Phoma*)	Seed treatment (10 g/kg seed)
8.	*T. harzianum* AZNF-5	Carrier based (1×10^8 cfu/g)	4 months at 55°C	Maru Sena 1	*Fusarium oxysporum* f. sp. *cumini* in cumin	Seed treatment (4 g/kg seed); soil application (1 kg/ ha with 50 kg FYM) before sowing
9.	*T. harzianum* (ICAR-CAZRI AZNF-5 and *Bacillus firmus* Bredemann and Werner 1933, ICAR-CAZRI AZ-1	Carrier based (1×10^8 cfu/g each)	4 months at 55°C	Mishrit Maru sena	*Macrophomina phaseolina* (Tassi) Goid. (1947) in legumes and oils seed crops	Seed treatment (10 g/kg seed with jiggery) and soil application (1 kg/ha with 40 kg FYM
10.	*B. bassiana* RF6	Talc (1×10^9 cfu/g)	8 months at 25-35°C	NRRI-BBLF	Rice leaf folder, *Cnaphalocrocis medinalis* (Guenée, 1854)	Foliar spray at 2 g/lit. of water
11.	*M. anisopliae* TF 19	Talc (1×10^9 cfu/g)	8 months at 25-35°C	NRRI-Malf	Rice leaf folder, *C. medinalis*	Foliar spray at 2 g/lit. of water
12.	*T. harzianum** Th4d	Liquid suspension concentrate (1×10^9 cfu/ml)	18 months at 25-35°C	Triguard Th-L	Phytophthora seedling blight, Macrophomina root rot and Fusarium wilt of safflower and gray mold of castor, Alternaria aster leaf blight and powdery mildew of sunflower	Seed treatment at 1 ml/kg seeds, foliar spray at 1-2 ml/ lit. of water

S.No.	Entomopathogenic fungi	Formulation	Shelf life	Trade name	Target pests	Dose
13.	*T. harzianum** Th4	Wettable powder (1×10^9 cfu/ml)	18 months at 25-35°C	Triguard Th-P	Phytophthora seedling blight, Macrophomina root rot; Fusarium wilt of safflower and castor, Aspergillus root rot in groundnut	Seed treatment at 10 g/kg of seeds
14.	*T. asperellum** Samuels, Lieckf. & Nirenberg 1999 Ta DOR 7316	Wettable powder (1×10^9 cfu/ml)	18 months at 25-35°C	Triguard Ta-P	Phytophthora seedling blight, Macrophomina root rot; Fusarium wilt of safflower and castor	Seed treatment at 10 g/kg seeds
15.	*B. bassiana* (ITCC 4513)	Liquid suspension concentrate (1×10^{12} cfu/ml)	24 months at 25-30°C	Mycoguard Bb-L	*Helicoverpa armigera* (Hubner) in pigeonpea	Two to three foliar sprays at 0.3 ml/lit of water at 10 days interval
16.	*T. harzianum** ICAR-IIHR Th-2	Wettable powder (2×10^6 cfu/g)	10 months at 25-35°C	Arka Krishi Vriddhi	*Meloidogyne incognita* (Kofoid & White, 1919), *Fusarium oxysporum* f. sp. *vasinfectum*, *S. rolfsii, F. solani* (Mart.) Sacc. (1881) in brinjal, tomato, carrot, okra	Seed treatment at 20 g/kg seed, nursery bed treatment at 50 g/m^2 for transplantable crops, soil application at 5 kg/ha after enrichment in 5 tons FYM before sowing or transplanting
17.	*T. viride** ICAR-IIHR Tv-5	Wettable powder (2 $\times 10^6$ cfu/g)	10 months at 25-35°C	Arka Krishi Veera	*M. incognita, Ralstonia solanacearum* (Smith 1896), *Erwinia carotovora* Winslow *et al.*, 1920, *Fusarium oxysporum* f. sp. *vasinfectum*, *Fusarium oxysporum* f. sp. *lycopersici*, *F.solani*	Seed treatment at 20 g/kg seed, nursery bed treatment at 50 g/m^2 for transplantable crops, soil application at 5 kg/ha after enrichment in 5 tons FYM before sowing or transplanting

S.No.	Entomopathogenic fungi	Formulation	Shelf life	Trade name	Target pests	Dose
18.	*Pochonia chlamydosporia** Zare and Gams IIHR-Vc-3	Carrier based (2×10^6 cfu/g)	10 months at 25-35°C	Arka Krishi Rakshak	*M. incognita* in brinjal, tomato, carrot, okra	Seed treatment at 20 g/kg seed, nursery bed treatment at 50 g/m^2 for transplantable crops, soil application at 5 kg/ha after enrichment in 5 tons FYM before sowing or transplanting
19.	*T. asperelloides asperelloides* 5R	Liquid formulation (5×10^{11} cfu/ml)	3 months at 25-35°C	Manjari Vineguard	Grapes powdery mildew	Soil drenching at 2 ml/lit. of water
20.	*T. afroharzianum*	Liquid (5×10^8 cfu/ml)	3 months at 25-35°C	Manjari Rakshak	Grapes powdery mildew	Foliar spray at 2 ml/lit. of water
21.	*T. harzianum* IARI P4	Wettable powder (10^8 cfu/g)	25 months at 25°C	Pusa 5 SD	*F. oxysporum* f. sp. *ciceris*, *S. rolfsii*, *S. sclerotiorum* (Lib.) de Bary (1884) in chickpea; *R. solani* J.G. Kühn 1858, *R. bataticola* (Taub.)Butl. in chickpea and mugbean, *F.* oxysporum f. sp. *lycopersici* in tomato, *P. ultimum*, *R. solani* in fresh bean, major seed borne fungal pathogens	Seed treatment at 4 g/kg of seeds
22.	*Purpureocillium lilacinum*	Wettable powder (2×10^6 cfu/ml)	10 months at 25-35°C	ARKA Krishi Kawach	*Meloidogyne incognita* in brinjal, tomato, carrot, okra	Seed treatment at 20 g/kg seed, nursery bed treatment at 50 g/m^2 for transplantable crops, soil application at 5 kg/ha after enrichment in 5 tonns FYM before transplanting or sowing

S.No.	Entomopathogenic fungi	Formulation	Shelf life	Trade name	Target pests	Dose
23.	*Bacillus thurigiensis* var. kurstaki	Liquid (1 × 10^8 cfu/ml)	12 months at 25-35°C	Shatpada Armour	Maize fall armyworm	Two to three foliar sprays at 10 ml/lit of water at 25, 35, 45 days after sowing
24.	*B. thuringiensis* var. kurstaki	Liquid (1 × 10^8 cfu/ml)	12 months at 25-35°C	Shatpada Terminator	*H. armigera*, *Plutella xylostella*, *Chilo partellus*, *Cnaphalocrocis medinalis*, *Leucinodes orbonalis*, *Amsacta albistriga*	Two to three foliar sprays at 20 ml/lit of water at pre-flowering and post flowering stages
25.	*Pseudomonas fluroescens*	Talc based (1 × 10^8 cfu/ml)	12 months at 25-35°C	Shatpada All Rounder	*Thrips* spp., in capsicum and Fusarium will of red gram	Foliar application at 20 g/lit of water at 20,30,40,50 days after transplanting for the management of thrips in capsicum; soil application in the root zone during 25,40,55 days after sowing at 2.5 kg/ha for management of red gram wilt; mix 2.5 kg of formulation in 250 kg farmyard manure and apply
26.	*B. albus*	Talc based (1 × 10^8 cfu/ml)	12 months at 25-35°C	Shatpada Master Blaster	*S. frugiperda, Tuta absoluta, Fusarium oxysporum* f. sp. *cucumerinum*	Foliar application at 20 g/lit of water at 20,30,40,50 days after sowing for fall armyworm and tomato pin worm; soil application in the root zone during 25,40,55 days after sowing at 2.5 kg/ha for management of cucumber wilt; mix 2.5 kg of formulation in 250 kg farmyard manure and apply

S.No.	Entomopathogenic fungi	Formulation	Shelf life	Trade name	Target pests	Dose
27.	*P. fluorescens*	Talc based (1×10^8 cfu/ml)	12 months at 25-35°C	Eco-Pesticide	Spot blotch of wheat, sheath blight of rice and wilt of tomato and chickpea	Seed treatment (10 g/kg seed)
28.	*B. firmus*	Carrier based (1×10^8 cfu/ml)	6 months at 25-35°C	Maru sena 3	*Macrophomina phaseolina* in legumes and oil seed	Seed treatment (30 g/kg seed with jiggery solution) before sowing
29.	*B. thuringiensis* var. kurstaki	Liquid suspension concentrate (1×10^{11} cfu/ml)	24 months at 25-35°C	Bioguard Bt-L	*Spodoptera litura* in soybean	Two foliar sprays at 3 ml/lit of water at 10 days interval.
30.	*P. fluorescens*	Wettable powder (2×10^8 cfu/ml)	10 months at 25-35°C	ARKA Krishi Samarakshak	*Meloidogyne incognita*, *Ralstonia solanacearum*, *Erwinia carotovora*, *Fusarium oxysporum* f. sp. *vasinfectum*, *Fusarium solani* in brinjal, tomato, carrot, okra	Seed treatment at 20 g/kg seed, nursery bed treatment at 50 g/m^2 for transplantable crops, soil application at 5 kg/ha after enrichment in 5 tonns FYM before transplanting or sowing
31.	*P. fluorescens*	Liquid (2×10^8 cfu/ml), carrier based (2×10^8 cfu/ml)	10-12 months at 25-35°C	ARKA krishi All Rounder and ARKA Plant Growth Booster	*Meloidogyne incognita*, *Fusarium oxysporum* f. sp. *vasinfectum*, *Fusarium solani* in several vegetable and fruit crops	Seed treatment at 20 g or 20 ml/kg seed, substrate treatmen with 10 ml or 10 g/kg of cocopeat, soil application at 5 kg or 5 li/ha after enrichment in 5 tons FYM before transplanting or sowing

*Licensed to private companies Source: Saxena et al. (2021)

Conclusions

Government regulations and the detrimental effects of chemical pesticides force a shift to alternate plant protection measures. As a result, microbial biopesticide, one of the environmentally friendly techniques, has become more significant in agriculture both globally and in India. Although a few factors, such as quality control and the identification of effective organisms, predispose the market and widespread use of biopesticide, central and state government initiatives, such as the establishment of assisted and non-aided biocontrol laboratories and intense R&D activities, support the growth of biopesticide steadily.

References

Dougoud J, Toepfer S, Bateman M, Jenner WH (2019) Efficacy of homemade botanical insecticides based on traditional knowledge. A review. Agron Sustainable Dev 39(4):37

Easwaramoorthy S, Jayaraj S (1987) Survey of granulosis virus infect in sugarcane borers, *Chilo infuscatellus* Snellen and *C. sacchariphagus indicus* (Kapur) in India. Int J Pest Manag 33(3):200-201.

Keswani C, Dilnashin H, Birla H, Birla H, Singh SP (2019) Regulatory barriers to agricultural research commercialization: a case study of biopesticides in India. Rhizosphere 11:100155. Doi.org/10.1016/j.rhishph.2019.100155

Keswani C, Sarma B, Singh H (2016) Synthesis of Policy Support, Quality Control, and Regulatory Management of Biopesticides in Sustainable Agriculture. 10.1007/978-981-10-2576-1_1.

Khergamer G (2019) Fall Armyworm attack: Maharashtra grappling with the chaos. Down to Earth.https://www.downtoearth.org.in/news/agriculture/fall-armyworm/attack-maharashtra- grappling-with-the-chaos-63400.

Koul O (2019) Nano-biopesticides today and future perspectives, Academic Press, Cambridge, 495p.

Kumar KK, Sridhar J, Murali-Baskaran RK, Senthil-Nathan S, Kaushal P, Dara SK, Arthurs S (2019) Microbial biopesticides for insect pest management in India: current status and future prospects. J Invert Pathol 165:74-81.

Martin PA, Hirose E, Aldrich JR (2007) Toxicity of *Chromobacterium subtsugae* to southern green stink bug (Heteroptera: Pentatomidae) and corn rootworm (Coleoptera: Chrysomelidae). J Econ Entomol 100(3):680-684.

Mishra J, Dutta V, Arora NK (2020) Biopesticides in India: technology and sustainability linkages. 3 Biotech 10:210. doi.org/10.1007/s13205-020-02192-7.

Pathak DV, Yadav R, Kumar M (2017) Microbial pesticides: Development, prospects and popularization in India, 455-471pp. In: DP Singh et al. (eds), Plant-microbe interactions in agro-ecological perspectives, Springer Nature, Singapore Pte Ltd

Sankaranarayanan C, Somasekhar N, Singaravelu B (2006) Biocontrol potential of entomopathogenic nematode *Heterorhabditis* and *Steinernema* against pupae and adults of white grub *Holotrichia serrata* F. Sugar Tech 8(4):268-271

Saxena AK, Chakdar H, Kumar M, Rajawat MVS, Dubey SC, Sharma TR (2021) ICAR Technologies: Biopesticides for Eco-friendly pest management. Indian Council of Agricultural Research, New Delhi, India 1-31pp.

Tikar S, Prakash S (2017) Fly ash-based *Bacillus thuringiensis israelensis* formulation: An ecofriendly approach. Indian J Med Res 146(6):680.

Wadke R (2018) Pink bollworm tears into the very fibre of Maharashtra's cotton growers. The Hindu Business Line. Https://www.thehindubuinessline.com/economy/agri-business/pink-bollworm-tears-into-the-very-fibre-of-maharashtras-cotton growers/article9969934.ece#. Access 22 Oct 2019

Links

http://ppqs.gov.in/divisions/cib-rc/biopesticide-registrant

https://ncof.dacnet.nic.in/Operational_Guidelines/Guidelines_for_Capital_Investment Subsidy.pdf

5

Nano-Biopesticides for Management of Insect Pests of Crops

Abstract

A type of pesticide known as a "biopesticide" is derived from natural sources such as plants, animals, minerals, and microbes. The employment of cutting-edge scientific tools will be extremely beneficial for enhancing the efficacy, greater applicability and adaptability, and storability of biopesticides. One such rapidly rising scientific discipline, nanotechnology, has considerable potential applications in agriculture, including the creation of carriers for pesticides, plant growth regulators, biofertilizers, nano-sensors, insecticides, food packaging materials, and gene transfer, among other things. Nanoparticles fall under the category of ultrafine particles and range in size from 1 to 100 nm. These nano-particles have special uses in various industries, including the production of nano-biopesticides. They differ from their bulk material due to their small size, shape, reactivity, and increased surface area to volume ratio. Without sacrificing on safety and health risks, nano-biopesticides offer higher accuracy in reaching their potential efficacy against target pests. This chapter covers several nano-biopesticides, their history, their use as biopesticides, modes of action, and their effectiveness against the pests they are intended to control.

Keywords: Nano-particle, Nano-biopesticide, AgNP, SiNP, Botanicals, Semiochemical

Introduction

Biopesticides are pesticides that are produced naturally by plants, animals, microbes, and other minerals. These represent less of a harm to humans and the environment than chemical insecticides. Nanotechnology is a rapidly developing scientific topic that has numerous uses in numerous industries, including agriculture. The transport of plant hormones, seed germination, water management, transfer of target genes, nano barcoding, nano sensors, and controlled release of agrichemicals are now being investigated as applications for nanotechnology in agriculture (Worrall et al. 2018). For their diverse applications in numerous industries, many scientists have modified nanoparticles for their size, shape, porosity, and/or surface tension, among other properties.

Nano-particles are a subcategory of ultrafine particles that range in size from 1 to 100 nm. These nano-particles have special uses in numerous industries due to their distinctive characteristics, which include small size, shape, reactivity, and a greater surface area to volume ratio. These nano-particles have numerous and varied uses in agriculture, including as pesticide transporters, plant growth regulators, biofertilizers, nano-sensors, insecticides, food packaging materials, and gene transfer agents, among others. Nano-particles give agricultural insect pest management techniques fresh dimensions. For targeted, controlled distribution of the active component in a biopesticide, nano-particles may be utilised. Compared to conventional biopesticides, nano-biopesticides provide a number of benefits.

Nano-Biopesticides

To increase their effectiveness, expand their potential applications, and solve numerous drawbacks, the various biopesticides, including microbial biopesticides, biochemical biopesticides, and plant-incorporated protectants, can be produced as nano-based biopesticides. In Table 1, a few of the nano-biopesticides were listed.

Table 1: Nanomaterials applied to bio-pesticide and their functions (An et al. 2022)

Bio-pesticide category	Active ingredients (AIs)	Material(s)	Function	Reference
Microbial pesticides (Bacterial pesticide)	*Bacillus thuringiensis* (*Bt*)	Nano-tubular sodium titanate	Effective in controlling cotton leafworm	Zaki et al. (2017)
Microbial pesticides (fungal pesti- cides)	*Beauveria bassiana* (Bals.)	Silica nano-particles and carbon fibers	Improving mortality to larvae of potato *Spodoptera litura*	Hersanti et al. (2020)
Plant-derived pesticides	Essential oils	Chitosan-coated nano-silver	Synergistic effect against a wide range of microorganisms	Gahukar et al. (2020)
Agricultural antibiotics	Avermectin	Poly (ethylene glycol)-carboxymethyl cellulose (PEG-CMC)	Improved the anti-UV ability & increased the biocompatibility of the Avermectin	Zhu et al. (2020)
	Validamycin and thifluzamide	Polylactic acid	The nano-particles prepared by compounding Validamycin with chemical pesticides showed better control of rice sheath blight, which was 4.2 times more effective than the control	Cui et al. (2020)
Biochemical pesticides	Gibberellic acid (GA)	Layered double hydroxides (LDH)	Promoted plant growth	Hafez et al. (2018)
Plant immunity elicitor-inducing antibacterial agents	Chitosan & Zinc oxide	Chitosan encapsulated zinc oxide nanocomposite	As an efficient biocompatible elicitor to improve agronomic traits of crops	Asgari-Targhi et al. (2021)

Botanical Based Nano-Biopesticides

Some of the nano-biopesticides are based on the botanical biopesticides such as essential oils, neem oil, neem powder etc., were found effective in manging many insect pests. The antifeedant and larvicidal activity of nano-biopesticide PONNEEM®-encapsulated tripolyphosphate cross-linked chitosan nanocarriers was reported against *Helicoverpa armigera* (Paulraj et al. 2017). Adel et al. (2018) tested the nano-emulsion of *Mentha piperita* essential oil against stored grain pest, *Tribolium castaneum* and reported that it was very much effective in controlling the pest with highest percent mortality in comparison to essential oil without nano-emulsion in wheat. Further, they also noticed that nano-emulsion significantly enhanced the germination percentage of wheat seeds. Similarly, Yang et al. (2009) reported the highest mortality (up to 80%) of *Tribolium castaneum* when treated with nano-biopesticides loaded with garlic essential oils. The list of nano-carriers used for biopesticides/active ingredients that target crop pests is provided in Table 2.

In an exclusive study, Palermo et al. (2021) tested the nano formulations prepared from eight commercial essential oils (*Pimpinella anisum*, *Artemisia vulgaris*, *Foenicum vulgare*, *Allium sativum*, *Lavandula angustifolia*, *Mentha piperita*, *Rosmarinus officinalis*, and *Salvia officinalis*) for their acute toxicity and repellence against confused flour beetle, *Tribolium confusum* and reported that all the nano-emulsions were found best repellent over time. The highest acute toxicity was noticed in garlic nano-emulsions with maximum mortality. Bidyarani and Kumar (2019) encapsulated the rotenone, a naturally occurring pesticide in the roots of Fabaceae plants, in zein nanoparticles by antisolvent precipitation method. They evaluated nano-encapsulated rotenone against plant pathogens *Pseudomonas syringae* and *Fusarium oxysporum* and reported excellent antimicrobial activity.

Table 2: List of nanocarriers for biopesticides/active ingredients that target crop pests (modified from Worrall et al. 2018)

Biopesticide	Nanoparticle	Crop	Target insect pest	Reference
Garlic essential oil	Polyethylene glycol	Rice (harvested)	Red flour beetle (*Triboleum castaneum*)	Yang et al. (2009)
Azadirachtin	Chitosan	-	Tobacco cutworm (*Spodoptera litura*) culture ovarian cell lines Sl-1	Lu et al. (2013)
Azadirachtin	Chitosan	-	-	Feng et al. (2012)
α-pinene and Linalool	Silica	Castor	Tobacco cutworm (*S. litura*), Castor semi-looper (*Achaea janata*)	Rani et al. (2014)
Abamectin	Porous silica	-	-	Wang et al. (2014)
Anacardic acid	Layered double hydroxide	Mustard	Tobacco cutworm (*S. litura*)	Nguyen et al. (2015)
Avermectin	Polydopamine	Cotton and corn	-	Jia et al. (2014)
Avermectin	Polydopamine	-	-	Sheng et al. (2015)
Suaeda maritima-based herbal coils	Silver	Cotton	Tobacco cutworm (*S. litura*)	Suresh et al. (2018)
PONNEEM	Chitosan	Cotton	Cotton bollworm (*Helicoverpa armigera*)	Paulraj et al. (2017)
Avermectin	Polydopamine	Cucumber and broccoli	Aphids	Liang et al. (2018)
Avermectin	Castor oil-based polyurethane	Corn leaves	-	Zhang et al. (2018)
Azadirachtin	Zinc oxide and chitosan	Groundnut	Bruchid (*Callosobruchus serratus*)	Jenne at al. (2018)
Carvacrol Linalool	Chitosan	-	Mite (*Tetranychus urticae*)	Campos et al. (2018)
Geraniol	Chitosan/Gum Arabic	-	Whitefly (*Bemicia tabaci*)	De Oliveira et al. (2018)

Biopesticide	Nanoparticle	Crop	Target insect pest	Reference
Satureja hortensis essential oil	Chitosan/TPP	-	Mite (*T. urticae*)	Ahmadi et al. (2018)
Geraniol and R-citronellal essential oils	Zein	-	Mite (*T. urticae*)	Oliveira et al. (2018)
Nicotine	Chitosan/TPP	-	House fly (*Musca domestica*)	Yang et al. (2018)
Carvacrol Linalool	Chitosan	Bean	Corn earworm (*H. armigera*), mites (*T. urticae*)	Campos et al. (2018)
Avermectin	porous hollow silica nanoparticles	*Brassica oleracea*	*Plutella xylostella* larvae	Kaziem et al. (2018)

Nano-formulations of Semiochemicals

One such naturally occurring semiochemical that is commonly employed to control insect pests is pheromones. They are somewhat unstable in nature as a result of isomerization, photooxidation, autooxidation, and volatility, among other processes (Deepa et al. 2013). In order to increase the effectiveness of pheromones in real-world settings, slow and controlled release formulations are crucial. Nano-formulations are the greatest options for delayed and controlled release of pheromones. By immobilising the pheromone into the nano gel, Deepa Bhagat et al. (2013) created a nano gel of methyl eugenol, a pheromone used to manage the fruit fly pest, *Bactrocera dorsalis*, and tested the gel's effectiveness in the field. Additionally, they discovered that pheromones based on nanogels were stable at room temperature and exhibited a reduced rate of evaporation, making handling and shipping simpler. When immobilised into nano-gels, methyl eugenol's shelf life was increased, and fruit fly pest trap catches were substantially higher than with methyl eugenol alone.

Abd El-Wahab et al. (2020) investigated the catchability of the red palm weevil, *Rhynchophorus ferrugineus*, and found that aggregation nano-gel pheromone traps attracted considerably more adult beetles than conventional pheromone traps in two seasons (55.33 and 46.33 adults/trap). White grubs (*Holotrichia consanguinea*) will be caught in ground nuts using a nano-gel formulation of the aggregation pheromone, methoxy benzene, created by Deep Bhagat et al. (2020). By immobilising the aggregation pheromone in a matrix and creating a viscoelastic semi-solid mass, a nanogel formulation was created. Additionally, they examined its effectiveness and discovered that the nanogel trap may capture up to 17.5 adult beetles daily.

Nano-Biopesticides Derived from Plants

Number of nano-biopesticides containing different nanoparticles were reported by many researchers. Plants serve as excellent sources for various nanoparticles which can be used as nano-biopesticides against various insect pests. The list of plant-derived nano-biopesticides, their application and nanoparticles present were presented in Table 3.

Table 3. Plant derived Nano-biopesticides and their efficacy against various insect pests (Krishnamurthy et al. 2020)

Plantname	Common name	NPs Present	Part of plant	Application	Reference
Acorus calamus	Sweet flag	Au, Ag	Root	Insecticide	Ganesan et al. (2015)
Agave americana	Sentry plant	Ag	Leaf	Whiteants	Ahmad et al. (2016)
Ageratum conyzoides	Chick weed	Ag	Leaf	Liceinhair	Wardani et al. (2019)
Albizialebbeck	Lebbeck	Ag, Ni, Fe	Seed, leaf, bark, root	Insecticide	Umar et al. (2010)
Albiziaprocera	White siris	Zn, Ag, Cu	Leaf	Insecticide	Jayakumarai et al. (2010)
Aloesecundiflora	New castle in chicken	Cd, Ag	Sap	Insecticide	Tippayawat et al. (2016)
Alysicarpus bupleurifolius	Alyce clover	Ag, P	Whole plant	Bed bugs, whiteants	Kasithevar et al. (2017)
Anacardium occidentale	Kaaju	Au, Ag, Cu, Pt	Shell oil	White ants, insecticide	Begum et al. (2018)
Anamirtacocculus	Indian berry	Zn, Ag, Au	Fruit	Insecticide	
Annona reticulate	Wild-sweetsop	Ag	Leaf, seed, bark	Body lice,insecticide	Parthiban et al. (2019)
Annona squamosa	Sugar apple	Ag, Fe	Seed, stem, bark, leaf, fruit	Body lice,insecticide	Vivek et al. (2010)
Arisaemator tuosum	Whipcord cobra lily	Ag, Au, Zn	Tubers	Insecticide	Kumar et al. (2018)
Aristolochia bracteolate	Worm killer	Ag	Juice	Insecticide	Doss (2015)
Artemisia japonica	Mug wort	Ag, Au, Fe	Whole plant	Insecticide, housefly repellent	Yu et al. (2019)
Artemisia nilagirica	Worm wood	Ag	Leaf	Insect repellent, prevent moths	Vijayakumar et al. (2013)
Azadirachta indica	Neem tree	Ag, Cu	Whole plant	Insecticide, rice and wheat weevil	Ahmad et al. (2016)
Bambusa arundinacea	Bambusa bambos	Ag	Shoot	Kill mosquito larvae	Kataria et al. (2017)
Biden spilosa	Black jack	Ag, Au	Leaves	Aphids	Kyomuhimbo et al. (2019)

Plantname	Common name	NPs Present	Part of plant	Application	Reference
Blumeae riantha	Buradi	Zn	Whole plant	Mosquito repellent	Benelli et al. (2017)
Boswellia serrata	Indian frankincense	Ag	Gum	Fumigation repel houseflies, mosquitoes	Kora et al. (2012)
Brassica campestris	Field mustard	Ag, Zn	Seed oil	Beetles	Khan et al. (2018)
Butea monosperma	Sacred tree	Ag, Au, Zn	Seed flower extract	White ants	Das et al. (2018)
Calotropis procera	Rubber bush	Ag, Zn, Ni, Fe	Leaf	Larvicidal	Gawade at al. (2017)
Canna bissativa	Kumbhi	Au, Ag	Whole plant leaf	Bugs & pests	Singh et al. (2018)
Capsicum annuum	Hot pepper	Cu, Ag, Au	Fruit, leaf	Thrips, aphids, white flies	Yuan et al. (2017)
Careya arborea	Karanda	Ag	Root, bark and leaf	Several	Nair et al. (2015)
Carica papaya	Papaya	Ag, Zn	Leaves, seeds	Several	Rathnasamy et al. (2017)
Carissa congesta	Hemp	Ag	Root and bark	Vet worms in wounds	Joshi et al. (2018)
Cassia hirsute	Cassia	Ag, Zn	Bark	Insecticide	Adesuji et al. (2016)
Cassytha filiformis	Love vine	Ag, Cu, Mg	Whole plant	Insecticide	Nasrollahzadeh et al. (2018)
Catunaregam spinosa	Mountain pomegranate	Sn, Zn, Ni	Fruit	Insecticide	Haritha et al. (2016)
Cinnamomum camphora	Camphor tree	Ag, Au, Pt, Pd	Bark powder	Protect clothes against insects	Huang et al. (2006)
Citrus limon	Lemon	Ag, Au	Dried leaf	Wheat weevil, flour beetle	Sujitha et al. (2013)
Commiphora wightii	Indian bdellium tree	Ag	Resin	Mosquito repellent	Sarkar (2017)
Cordia latifolia	Sebestan plum	Ag	Leaves	Maize weevil, butterfly	Ioset et al. (2000)

Plantname	Common name	NPs Present	Part of plant	Application	Reference
Coryphaum braculifera	Talipot palm	Cu, Ag	Young fruit	Insect repellent	Abdel-Wahab et al. (2019)
Croton roxburghii	Croton	Ag	Seed	Insecticide	Panda et al. (2010)
Cucumis melo	Muskmelon	Ag	Leaf	lice	Haryani et al. (2018)
Cucumis sativus	Cucumber	Ag, Cu, Zn	Rhizome	lice and insects	Zhao et al. (2014)
Curcuma longa	Turmeric	Ag, Zn	Rhizome	Drive away ants	Shameli et al. (2012)
Cuscuta reflexa	Amarbel	Cu, Ag	Whole plant	lice	Naghdi et al. (2018)
Cymbopogan nardus	Citronella grass	Ag	Whole plant	Mosquito repellent	Kamarudin et al. (2019)
Derriss candens	Gewel vine	Ag	leaf, bark	Insecticide	Firdhouse et al. (2013)
Derris trifoliate	Karanjvel	Ag	Bark	Insecticide	Kumar et al. (2017)
Desmodium triflorum	Tick clover	Ag, Cu, Au	Whole plant	Insecticide	Ahmad et al. (2011)
Dioscorea hispida	Asia ticbitteryam	Ag	Bark	Insecticide	Ashri et al. (2014)
Duranta erecta	Brazili ansky flower	Ag, Zn	Whole plant	Insecticide	Ravindran et al. (2016)
Euphorbia antiquorum	Spurge	Ag	Milky juice	Maggot sin wound	Rajkuberan et al. (2017)
Euphorbia dracunculoides	Dragon spurge	Ag	Latex	Kill slice	Annamalai et al. (2013)
Euphorbia thymifolia	Thyme leaf	Pd	Whole plant	Flies, mosquitoes	Nasrollahzadeh et al. (2016)
Fioria vitifolia	Grape leaved mallow	Ag	Root, bark	Kill slice	Ghosh et al. (2015)

Plantname	Common name	NPs Present	Part of plant	Application	Reference
Glorio sasuperb	Flame lily	Ag, Au, Ce, Cu ,Pt, Pd	Leaf	Liceinthe hair	Ashokkumar et al. (2013)
Haldina cordifolia	Kadam	Ag	Bark	Insecticide	Khan et al. (2019)
Hard wickiabinate	Anjan	Zn	Wood	Insecticide	Gunaselvi et al. (2010)
Harpullia arborea	Tulip wood	Zn, Sn	Bark	Leech repellent	Mohan et al. (2018)
Holarrhena pubescens	Indrajao	Ag	Flower, seed	Insecticides	Venkata Subbaiah et al. (2013)
Hyptissua veolens	American mint	Ag, Cu	Twig	Repel bed bugs	Elumalai et al. (2017)
Kalanchoe integra	Never die	Au, Cu	Leaf	Insecticide	Patel et al. (2019)
Lagenandra ovate	Malayans word	Ag	Whole plant	Insecticide	Bokaeian et al. (2015)
Lavandula bipinnata	Lavender	Zn	Whole plant	Insect repellent	Shaikh et al. (2014)
Lavendulalawii	Lavender	Ag	Wholeplant	Insect repellent	Kulkarni et al. (2013)
Leonotis nepetifolia	Klip dagga	Ag	Leaf	House fly repellent	Al-Sheddi et al. (2018)
Leucas aspera	Thumbai	Ag, Ce, Cu	Whole plant	Insecticide	Malleshappa et al. (2015)
Lippia javanica	Fever tea	Ag	Leaf	Insecticide	Kumar et al. (2016)
Madhu calongifolia	Ilippai	Ag, Cu, Au	Seed, seed, oil, cake	Worm killer, insect repellent	Sharma et al. (2019)
Melaleuca leucadendron	Caju puttree	Au, Ag	Oil	Mosquito repellent	Souza et al. (2017)
Melia azadarach	Chinaberry	Ag	Fruit, seed	Insecticide	Anbu et al. (2016)

Plantname	Common name	NPs Present	Part of plant	Application	Reference
Melia volkensii	Melia	Ag	Fruit pulp	Termites	Kamau et al. (2016)
Millettia extensa	Benth	Ag	Root	Insecticide	Panda et al. (20160
Mimosa pudica	Shame plant	Ag, Zn, Fe, Au, Cu	Leaf	Veterinary wound maggots	Fatimah et al. (2016)
Mundulea sericea	Cork bush	Zn	Seed, root, bark	Insecticide	Chaithong et al. (2006)
Nigella sativa	Black seed	Ag, Zn	Seed	Pesticide	Amooaghaie et al. (2015)
Ocimum americanum	Hoary basil	Ag	Whole plant	Insecticide	Anuradha et al. (2014)
Ocimum gratissimum	Ramtulsi	Ag, Au	Whole plant	Insect repellent	Das et al. (2017)
Ocimum kilimandscharicum	Camphor Basil	Ag	Leaves, flower	Mosquito, fleas	Selvarani et al. (2016)
Ocimum tenuiflorum	Tulsi	Ag	Whole plant	Insect repellent	Singhal et al. (2011)
Peganum harmala	Wild rue	Ag, Zn	Root	Mosquito repellent	Fazlzadeh et al. (2017)
Psidia punctulata	Mpepe	Ti, Cu	Leaves	Lice, fleas, mites	Zinjarde et al. (2011)
Pongamia pinnata	Indian beech	Ag	Seed, root, seed oil	Repellent, insecticide	Paul et al. (2018)
Riccinus communis	Castor bean	Ag, Au	Seed oil	Flies repellent, rice moth, rice weevil	Ojha et al. (2017)
Ruta graveolens	Herbof grace	Zn, Ag	Whole plant	Insects	Lingaraju et al. (2016)
Sarcostm maviminale	Caustic vine	Ag	Leaf	White ants	Kannan et al. (2018)
Securidacalon gepedunculata	Violet tree	Ag	Whole plant	Stored grain pest	Ojewole et al. (2008)

Plantname	Common name	NPs Present	Part of plant	Application	Reference
Senna didymobotrya	Popcorn senna	Ag	Leaves	Nematodes	Vijayakumari et al. (2018)
Solanum nigrum	Sodom apple	Ag	Fruit, leaf	Insecticides	Venkat Kumar et al. (2017)
Strychno snuxvomica	Poison nut	Zn, Au, Ag	Fruit, seed	White ants	Steffy et al. (2018)
Strychnos spinosa	Monkey orange	Ag	Whole plant	Insecticides	Isa et al. (2014)
Symplytum officinale.	Common comfrey	Ag	Leaf, root	Insecticides	Singh et al. (2018)
Tagetes minuta	Wild marigold	Ag	Leaf, flower	Insecticides	Shahzadi et al. (2015)
Tanacetum cinerariifolium	Pyrethrum	Ag, Au	Flower	Bees and insects	Kitherian et al. (2016)
Tephrosia purpurea	Fish poison	Ag, Au	Wood, roots	Cotton and woollen cloth moths	Srikar et al (2016)
Tithonia diversifolia	Marigold	Ag	Leaf, flower	Insecticides	Tran et al. (2013)
Trachylo biumammi	Ajwain	Ag, Zn	Seed	Mosquito repellent	Chouhan et al. (2017)
Trigonella foenumgraecum	Fenugreek	Au, Ag	Seed	Insect repellent	Aswathy et al. (2012)
Vernonia amygdalina	Bitter leaf	Ag	Leaf	Insecticides	Widyaningtyas et al. (2019)
Vernonia anthelminticum	Kalijiri	Zn, Au	Seed	Fleas	Karthikeyan et al. (2008)
Vitex negundo	Chaste tree	Ag, Zn	Leaf	Insect repellent	Prabhu et al. (2013)
Vitex trifolia	Arabianlilac	Ag, Zn, Au	Leaf	Insect repellent	Elumalai et al. (2015)

Nanoparticles as Nano-Biopesticides

With the growing advancement of science and awareness about environment and pesticide free food materials, eco-friendly measures are employed for insect pest management practices. Nanotechnology offers one such solution to eco-friendly control of insect pests in the form of nano-particles. Lot of evidence proved that many nano-particles offers as insecticide/acaricide against a range of insect pests across the genera. Some of the nano-particles used as biopesticides are listed in Table 4.

Table 4: Mode of action of nano-particles as biopesticides (Benelli, 2018)

Tested nanomaterial (dose or concentration)	Insect target	Morphological damages and/ or mode of action	References
Various green and microbial synthesized Ag, Au, and ZnO nanoparticles	*Aedes aegypti, Anopheles stephensi*	Midgut, epithelial cell, and cortex damages, with accumulation of nanoparticles in the midgut. Shrinkage in the abdominal region, thorax shape changes, midgut damages, loss of lateral hairs, anal gills and brushes	Banumathi et al. (2017); Kalimuthu et al. (2017), Sundararajan and Kumari (2017), Abinaya et al. (2018), Ishwarya et al. (2018)
Ag nanoparticles preparedusing Cassia fistula extract(LC_{50} = 3.6 and 1.7 mg/l, respectively)	*Aedes albopictus, Culex pipiens pallens*	4th instar larvae showed a decrease of total protein levels; nano Ag also reduced acetylcholinesterase and α- and ß-carboxylesterase activities	Fouad et al. (2018)
Ag nanoparticles fabricated using salicylic acid and 3,5-initrosalicylic acid(1–12 ppm)	*Aedes albopictus*	4th instar larvae showed a decrease of total proteins, esterase, acetylcholine esterase, and phosphatase enzymes.	Ga'al et al. (2018)
Ag nanoparticles (0.2, 0.5, and 1 mg/l)	*Chironomus riparius*	GST genes up- or down regulated, according to tested concentration and duration of exposure, highest mRNA expression was in delta3,Sigma4 and Epsilon1 GST class	Nair and Choi (2011)

Tested nanomaterial (dose or concentration)	Insect target	Morphological damages and/ or mode of action	References
Ag nanoparticles (up to 4 mg/l)	*Chironomus riparius*	Down regulation of the ribosomal protein gene (CrL15) regulating ribosomal assembly, thus protein synthesis. Up regulation of gonadotrophin releasing hormone gene (CrGnRH1) and Balbiani ring protein gene (CrBR2.2), which can indicate the activation of gonadotrophin releasing hormone mediated signal transduction pathways and reproductive failure.	Nair et al. (2011)
Ag nanoparticles (0.2, 0.5, and 1 mg/l)	*Chironomus riparius*	Expression of the ecdysone receptor gene was up or Down regulated according to the exposure time	Nair and Choi (2012)
Ag nanoparticles (0.2, 0.5, and 1 mg/l)	*Chironomus riparius*	Up regulation of Mn superoxide dismutase; transcript levels of catalase, phospholipidhydroperoxide glutathione peroxidase 1 and thioredoxin reductase 1 upregulated. Boosted expression of Delta-3, sigma-4, and epsilon-1 classes of glutathione S-transferases	Nair et al. (2013)
Ag nanoparticles (< 50 mg/l)	*Drosophila melanogaster*	Loss of melanin cuticular pigments, reduced vertical flight ability, reduced activity of Cu-dependent enzymes (tyrosinase and Cu-Zn superoxide dismutase); nanoAg coupled with membrane-bound Cu transporter proteins lead sequestration of Cu, mimicking Cu starvation	Armstrong et al. (2013)

Tested nanomaterial (dose or concentration)	Insect target	Morphological damages and/ or mode of action	References
Ag nanoparticles (25–50 μg/ml nanoAg 4.7 nm and250–1000 μg/ml nanoAg 42 nm)	*Drosophila melanogaster*	Lack of mutagenic and recombinogenic activity. However, both nano-Ag 4.7 and 42 nm evoked pigmentation defects and locomotor ability decrease in adult flies	Ávalos et al. (2015)
Ag nanoparticles (10–50 μg/ml)	*Drosophila melanogaster*	Accumulation of reactive oxygen species (ROS) in the fly tissues leading to ROS-mediated apoptosis, DNA damage, and autophagy; activation of the Nrf2-dependent antioxidant pathway	Mao et al. (2018)
Ag and TiO_2 nanoparticles (0.005 to 0.05%)	*Drosophila melanogaster*	Progeny loss and a decrease in developmental success	Philbrook et al. (2011)
Ag nanoparticles (500 to4000 mg/l)	*Spodoptera litura* and *Achaea janata*	Nano-induced oxidative stress in moth larval guts, with enhanced antioxidant enzyme levels	Yasur and Usha-Rani (2015)
Ag nanoparticles synthesized using the *Punica granatum*peel extract (LC_{50} = 19.21 μg/larva)	*Spodoptera litura*	Reduction of amylase, protease, lipase, and invertase activities; gut microflora and the extracellular enzyme production decreased, along with weight, pH, and total heterotrophic bacterial population	Bharani and Namasivayam (2017)
Nanostructured Al_2O_3 (60–500 ppm)	*Sitophilus oryzae*	Bind to the beetle cuticle due to triboelectric forces, sorbing its wax layer by surface area phenomena, resulting in insect dehydration	Stadler et al. (2017)
Au nanoparticles (87.44 μg/gin the diet)	*Blattella germanica*	Disrupted reproduction and development	Small et al. (2016)
Au nanoparticles fabricated using latex of *Jatropha curcas* (500–1000 μl)	*Aedes aegypti*, beetles, and Mealy bugs	Triggered trypsin inhibition	Patil et al. (2016)

Tested nanomaterial (dose or concentration)	Insect target	Morphological damages and/ or mode of action	References
Carbon black and multiwalled nanotubes (3.3 and 3.1 mg, respectively)	*Drosophila melanogaster*	Strong adherence of the nanomaterials to the fly body parts, leading to impaired motor functions and insect mortality	Liu et al. (2009)
Graphene oxide nanoparticles (0.1 μl per 100 mg ofinsect's body weight)	*Acheta domesticus*	Increased enzymatic activity of catalase and glutathione peroxidases, as well as heat shock protein (HSP 70) and total antioxidant capacity levels	Dziewięcka et al. (2016)
Carbon-dot-Ag nanohybrid(LC_{50} values from 0.30 to0.76 ppm)	*Anopheles stephensi*, *Culex quinquefasciatus*	Deformation of larval body, presence of Ag(2.93%) in the tissues of treated mosquitoes, cuticle, and cellular organization damages	Sultana et al. (2018)
Polystyrene nanoparticles (20–500 μg/ml)	Insect cells (BACULOSOMES®)	Inhibited the enzymatic activity of CYP450 isoenzymes in BACULOSOMES®	Fröhlich et al. (2010)
SiO_2 nanoparticles	Different species, with special reference to stored product pests	Physio-sorbed by the insect cuticular lipids, causing major damages, followed by the insect's death	Barik et al. (2008), Debnath et al.(2011), Athanassiou et al. (2018)
SiO_2nanoparticles (LudoxTMA) (≥ 34 mg/l)	*Bombus terrestris*	Midgut epithelial injury in intoxicated workers	Mommaerts et al. (2012)
TiO_2nanoparticles (5 μg/ml)	*Bombyx mori*	Upregulation of pi3k and P70S6K [rapamycin (TOR) signalling pathway]; 4 cytochrome P450genes (20-hydroxyecdysone biosynthesis),were up-regulated; 20-hydroxyecdysone biosynthesiswas stimulated; reduced development and moulting duration were noted	Li et al. (2014)

Ag Nanoparticles (AgNPs)

Many researchers studied the green synthesis of AgNPs and their efficacy in controlling many agricultural and household insect pests. Devi et al. (2014) synthesized AgNPs from leaf aqueous extract of *Euphorbia hirta* L. (Malpighiales: Euphorbiaceae) and tested against larvae and pupae of cotton bollworm, *H. armigera* and observed the susceptibility of all stages. Similarly, Marimuthu et al. (2011) synthesized AgNPs from leaf aqueous extract of *Mimosa pudica* L. (Fabales: Fabaceae) and tested on larvae of mosquitos *C. quinquefasciatus* and *A. subpictus* and larvae of the tick *Rhipicephalus microplus* Canestrini (Acari: Ixodidae) and found their susceptibility to AgNPs. Kantrao et al. (2017) synthesized AgNPs from leaf extracts of the Peepal tree, *Ficus religiosa* and the banyan tree, *Ficus benghalensis* and tested on *H. armigera* and found that AgNPs modulated gut protease activity in larvae of *H. armigera*. Vinayagamoorthi et al. (2015) synthesized AgNPs from aqueous extract of *Sargassummuticum* (Yendo) Fensholt (Fucales: Sargassaceae) and tested on 4th instar larvae of the common castor, *Ariadne merione* (Cramer) (Lepidoptera: Nymphalidae) and observed the physiological and anatomical abnormalities in the larval body.

Silica Nanoparticles (SiNPs)

SiNPs are most studied nanoparticles either as nanocarriers or as biopesticide in one or the other form (fungicide, bactericide, pheromone, plant growth regulator) against number of insect pests both under field and storage conditions (Barik et al. 2008). The mode of action of SiNPs is similar to that of bulk silica where in SiNPs are physio-sorbed by the cuticular lipids destroying the protective barrier and thereby causing insect to death. World Health Organization (WHO) declared use of amorphous silica as nano-biopesticide is safe to humans (Athanassiou et al. 2017). The surface charged SiNPs (3-5 nm) were successfully used for management of insect pests agricultural and veterinary importance across the taxa (Ulrichs et al. 2005). Debnath et al. (2011) reported the application of SiNPs caused 100% mortality of adults of storage pest rice, rice weevil *Sitophilus oryzae* (L.) (Coleoptera: Curculionidae). Fouad et al. (2016) reported that application of SiNPs (600 ppm) along with Jasmonic acid at rate 1.141 μM/plant significantly reduced the tomato fruit damage by *Tuta absoluta* larvae.

El-Samahy et al. (2015) reported that 70.11 and 60.56% reduction in larvae of *Spodoptera littoralis* (75 and 60 g/fed) in sugarbeet due to application of SiNPs. Similarly, El-Helaly et al. (2016) tested the SiNPs at 200, 300, 400 and 500 ppm along with bulk silica and diazinon against *S. littoralis* in squash and reported that 73.07, 79, 72, 87.88 and 89.82% mortality of larvae at respective doses. Shoaib et al. (2018) tested the SiNPs against *P. xylostella* larvae @ 1 mg cm^{-2} and reported that mortality percentage increased from 58% and 85% at 24 and 72 h after treatment and further noticed that the larval death was due to desiccation, body wall abrasion, and spiracle blockage.

Conclusions

Nanotechnology offers grater applicability in various fields of agriculture and allied scineces. It has tremendous role in insect pest management strategies such as nano carrier, nano emulsions, nano pesticides and so on. Biopesticides are one such class of compounds which offer environmentally friendly, residual and pollution free control of insect pests in various crops. Nano form of biopesticides offer greater advantage over traditional biopesticides in achieving maximum potential in controlling target insect pests. More studies are required on effect of nanobiopesticides on natural enemies and their role in tri-trophic interactions. With this, it can be concluded that nanobiopestcodes are the best alternatives for chemical pesticides and traditional biopesticides for successful and sustainable management of insect pests of various crops.

References

Abd El-Wahab AS, Abd El-Fattah AY, El-Shafei WKM, El Helaly AA (2020) Efficacy of aggregation nano gel pheromone traps on the catchability of *Rhynchophorus ferrugineus* (Olivier) in Egypt. Brazilian J Biol 81:452-460

Abdel-Wahab DA, Othman NARM, Hamada AM (2019) Effects of copper oxide nanoparticles to *Solanum nigrum* and its potential for phytoremediation. Plant Cell Tiss Organ Cult 137:525-539

Abinaya M, Vaseeharan B, Divya M, Sharmili A, Govindarajan M, Alharbi NS, Kadaikunnan S, Khaled JM, Benelli G (2018) Bacterial exopolysaccharide (EPS)-coated ZnO nanoparticles showed high antibiofilm activity and larvicidal toxicity against malaria and Zika virus vectors. J Trace Elem Med Biol 45:93-103

Adel MM, Massoud MA, Mohamed MIE, Abdel-Rheim KH, El-Naby SSA (2018) New Nano-biopesticide Formulation of *Mentha piperita* L. (Lamiaceae) Essential Oil Against Stored Product Red Flour Beetle *Tribolium castaneum* Herbs and its Effect on Storage. Adv Crop Sci Tech 6:409

Adesuji ET, Oluwaniyi OO, Adegoke HI, Moodley R, Labulo AH, Bodede OS, Oseghale CO (2016) Investigation of the larvicidal potential of silver nanoparticles against *Culex* quinquefasciatus: A case of a ubiquitous weed as a useful bioresource. J Nanomater 2016:1-11

Ahmad B, Khan I, Shireen F, Bashir S, Azam S (2016) Green synthesis, characterisation and biological evaluation of AgNPs using *Agave americana, Mentha spicata* and *Mangifera indica* aqueous leaves extract. IET Nanobiotechnol 10:281-287

Ahmad N, Sharma S, Singh VN, Shamsi SF, Fatma A, Mehta BR (2011) Biosynthesis of silver nanoparticles from *Desmodium triflorum:* A novel approach towards weed utilization. Biotechnol Res Int 2011.

Ahmadi Z, Saber M, Akbari A, Mahdavinia GR (2018) Encapsulation of *Satureja hortensis* L. (Lamiaceae)in chitosan/TPP nanoparticles with enhanced acaricide activity against *Tetranychus urticae* Koch (Acari:Tetranychidae). Ecotoxicol Environ Saf 161:111-119

Ahmed S, Saifullah Ahmad M, Swami B, Ikram S (2016) Green synthesis of silver nanoparticles using *Azadirachta indica* aqueous leaf extract. J Radiat Res Appl Sci 9: 1-7

Al-Sheddi ES, Farshori NN, Al-Oqail MM, Al-Massarani SM, Saquib Q, Wahab R, Musarrat J, Al-Khedhairy AA, Siddiqui MA (2018) Anticancer potential of green synthesized silver nanoparticles using extract of *Nepeta deflersiana* against human cervical cancer cells (HeLA). Bioinorg Chem Appl 2018

Amooaghaie R, Saeri MR, Azizi M (2015) Synthesis, characterization and biocompatibility of silver nanoparticles synthesized from *Nigella sativa* leaf extract in comparison with chemical silver nanoparticles. Ecotoxicol Environ Safety 120:400-408.

An C, Sun C, Li N, HuangB, Jiang J, Shen Y, Wang C, Zhao X, Cui B, Wang C, LiX (2022) Nanomaterials and nanotechnology for the delivery of agrochemicals: strategies towards sustainable agriculture. J Nanobiotechnol 20(1):1-19

Anbu P, MuruganK, Madhiyazhagan P, Dinesh D, Subramaniam J, Panneerselvam C, Suresh U, Alarfaj AA, Munusamy MA, Higuchi A, Hwang JS, Kumar S, Nicoletti M, Benelli G (2016) Green-synthesised nanoparticles from *Melia azedarach* seeds and the cyclopoid crustacean *Cyclops vernalis*: an eco-friendly route to control the malaria vector *Anopheles stephensi*? Nat Prod Res 30:2077-2084

Annamalai A, Christina VL, Sudha D, Kalpana M, Lakshmi PT (2013) Green synthesis, characterization and antimicrobial activity of Au NPs using *Euphorbia hirta* L. leaf extract. Colloids Surf B Biointerfaces 108:60-65

Anuradha G, SyamaB, Ramana MV (2014) *Ocimum americanum* L. leaf extract mediated synthesis of silver nano particles: A noval approach towards weed utilisation. Arch Appl Sci Res 6:59-64

Armstrong N, Ramamoorthy M, Lyon D, Jones K, Duttaroy A (2013) Mechanism of silver nanoparticles action on insect pigmentation reveals intervention of copper homeostasis. PLoS One 8(1):e53186.

Asgari-Targhi G, Iranbakhsh A, Ardebili ZO, Tooski AH (2021)Synthesis and characterization of chitosan encapsulated zinc oxide (ZnO) nanocomposite and its biological assessment in pepper (*Capsicum annuum*) as an elicitor for in vitro tissue culture applications. Int J Biol Macromol 189:170-82

Ashokkumar S, Ravi S, Velmurugan S (2013) Green synthesis of silver nanoparticles from *Gloriosa superba* L. leaf extract and their catalytic activity. Spectrochim Acta Part A Mol Biomol Spectrosc 115:388-392

Ashri A, Yusof M, Jamil M, Abdullah A, Yusoff S, Arip M, Lazim A (2014) Physicochemical characterization of starch extracted from malaysian wild yam (*Dioscorea hispida* dennst.). Emirates J Food Agric 26:652-658

Aswathy Aromal S, Philip D (2012) Green synthesis of gold nanoparticles using *Trigonella foenum-graecum* and its size-dependent catalytic activity. Spectrochim Acta Part A Mol Biomol Spectrosc 97:1-5

Athanassiou CG, KavallieratosNG, Benelli G, Losic D, Usha Rani P, Desneux N (2018) Nanoparticles for pest control: current status and futureperspectives. JPest Sci 91(1):1-15

Ávalos A, Haza AI, Drosopoulou E, Mavragani-Tsipidou P, Morales P (2015) In vivo genotoxicity assessment of silver nanoparticles of different sizes by the Somatic Mutation and Recombination Test (SMART) on *Drosophila*. Food Chem Toxicol 85:114-119

Balavigneswaran CK, Sujin Jeba Kumar T, Moses Packiaraj R, Prakash S (2014) Rapid detection of Cr(VI) by AgNPs probe produced by *Anacardium occidentale* fresh leaf extracts. Appl Nanosci 4:367-378

Banumathi B, Vaseeharan B, Ramachandran I, Marimuthu Govindarajan M, Alharbi NS, Kadaikunnan S, Khaled JM, Benelli G (2017) Toxicity of herbal extracts used in ethno-veterinary medicine and green encapsulated ZnO nanoparticles against *Aedes aegypti* and microbial pathogens. Parasitol Res 116:1637-1651

Barik TK, Sahu B, Swain V (2008) Nanosilica-from medicine to pest control. Parasitol Res 103:253-258

BegumSR, RaoDM, Reddy PDS (2018) Role of green route synthesized silver nanoparticles in medicinal applications with special reference to cancer therapy. Biosci Biotech Res Asia15

BenelliG, Govindarajan M, Rajeswary M, Senthilmurugan S, Vijayan P, Alharbi NS, Kadaikunnan S, Khaled JM (2017) Larvicidal activity of *Blumea eriantha* essential oil and its components against six mosquito species, including Zika virus vectors: the promising potential of (4E,6Z)-allo-ocimene, carvotanacetone and dodecyl acetate. Parasitol Res116:1175-1188

Bhagat D, Samanta SK, Bhattacharya S (2013) Efficient management of fruit pests by pheromone nanogels. Sci Rep 3(1):1-8

Bharani RA, Namasivayam SKR (2017). Biogenic silver nanoparticles mediated stress on developmental period and gut physiology of major lepidopteran pest *Spodoptera litura* (Fab.)(Lepidoptera: Noctuidae)-An eco-friendly approach of insect pest control. J Environ Chem Eng 5(1):453-467

Bidyarani N, Kumar U (2019) Synthesis of rotenone loaded zein nano-formulation for plant protection against pathogenic microbes. RSC Adv 9(70):40819-40826

Bokaeian M, FakheriBA, Mohasseli T, Saeidi S (2015) Antibacterial activity of silver nanoparticles produced by *Plantago Ovata* seed extract against antibiotic resistant *Staphylococcus aureus*.Int J Infect 2:e22854

Campos EV, Proença PL, Oliveira JL, Melville CC, Vechia JF, Andrade DJ, Fraceto LF (2018) Chitosannanoparticles functionalized with-cyclodextrin: A promising carrier for botanical pesticides. Sci Rep 8:2067

Chaithong U, Choochote W, Kamsuk K, Jitpakdi A, Tippawangkosol P, Chaiyasit D, Champakaew D, Tuetun B, Pitasawat B (2006) Larvicidal effect of pepper plants on *Aedes aegypti* (L.) (Diptera: Culicidae). J Vector Ecol 31:138-144

ChouhanN, Ameta R, Meena RK (2017) Biogenic silver nanoparticles from *Trachyspermum ammi* (Ajwain) seeds extract for catalytic reduction of p -nitrophenol to p -aminophenol in excess of NaBH 4. J Mol Liq 230:74-84

Cui J, Sun C, Wang A, Wang Y, Zhu H, Shen Y, Li N, Zhao X, Cui B, Wang C, Gao F, Zeng Z, Cui H (2020) Dual-functionalized pesticide nanocapsule delivery system with improved spreading behavior and enhanced bioactivity. Nanomater 10(2):220

Das B, Dash SK, Mandal D, Ghosh T, Chattopadhyay S, TripathyS, Das S, Dey SK, Das D, Roy S (2017) Green synthesized silver nanoparticles destroy multidrug resistant bacteria via reactive oxygen species mediated membrane damage. Arabian J Chem 10:862-876

Das M, Smita SS (2018) Biosynthesis of silver nanoparticles using bark extracts of *Butea monosperma* (Lam.) Taub. and study of their antimicrobial activity. Appl Nanosci 8:1059-1067

De Oliveira JL, Campos EVR, Pereira AES, Nunes LE, da Silva CC, Pasquoto T, Lima R,Smaniotto G, Polanczyk RA, FracetoLF (2018) Geraniol encapsulated in chitosan/ gum arabic nanoparticles: Apromising system for pest management in sustainable agriculture. J Agric Food Chem 66:5325-5334

Debnath N, Das S, Seth D, Chandra R, Bhattacharya SC, Goswami A (2011) Entomotoxic effect of silica nanoparticles against *Sitophilus oryzae* (L.) J Pest Sci 84:99-105

Deepa Bhagat, Jakhar BL, Subaharan K, Bakhtavatsalam N, Saini KK, Baloda AS, Sreedevi K (2020) Slow release pheromone formulations for the management of *Holotrichia consanguinea* (Blanchard)-laboratory to market. ICAR-NBAIR publication 24/2020

Devi GD, Murugan K, Selvam CP (2014) Green synthesis of silver nanoparticles using *Euphorbia hirta* (Euphorbiaceae) leaf extract against crop pest of cotton bollworm, *Helicoverpa armigera* (Lepidoptera: Noctuidae). J Biopest 7:54-66

DossA (2015) Biosynthesis and characterization of nanoparticles from three *Aristolochia species*. World J Pharm Pharma Sci 4:1094-1104

Dziewięcka M, Karpeta-Kaczmarek J, Augustyniak M, Majchrzycki Ł, Augustyniak-Jabłokow MA (2016) Evaluation of in vivo graphene oxide toxicity for *Acheta domesticus* in relation to nanomaterial purity and time passed from the exposure. J Hazard Mat 305:30-40

El-Helaly AA, El-Bendary HM, Abdel-Wahab AS, El-Sheikh MAK, Elnagar S (2016) The silica-nano particles treatment of squash foliage and survival and development of *Spodoptera littoralis* (Bosid.) larvae. Pest Control 5:6

El-Samahy MFM, Khafagy IF, El-Ghobary AMA (2015) Efficiency of silica nanoparticles, two bioinsecticides, peppermint extract and insecticide in controlling cotton leafworm, *Spodoptera littoralis* Boisd and their effects on some associated natural enemies in sugar beet fields. Mansoura J Plant Prot Pathol 6:1221-1230

Elumalai D, Hemavathi M, Deepaa CV, Kaleena PK (2017) Evaluation of phytosynthesised silver nanoparticles from leaf extracts of *Leucas aspera* and *Hyptis suaveolens* and their larvicidal activity against malaria, dengue and filariasis vectors. Parasite Epidemiol Control 2:15-26

Elumalai K, Velmurugan S, Ravi S, Kathiravan V, Raj GA (2015) Bio-approach: Plant mediated synthesis of ZnO nanoparticles and their catalytic reduction of methylene blue and antimicrobial activity. Adv Powder Technol 26:1639-1651

Fatimah I, Pradita RY, Nurfalinda A (2016) Plant extract mediated of ZnO nanoparticles by using ethanol extract of *Mimosa Pudica* leaves and coffee powder. Procedia Eng148:43-48

Fazlzadeh M, Khosravi R, Zarei A (2017) Green synthesis of zinc oxide nanoparticles using *Peganum harmala* seed extract, and loaded on *Peganum harmala* seed powdered activated carbon as new adsorbent for removal of Cr(VI) from aqueous solution. Ecol Eng 103:180-190

Feng BH, Peng LF (2012) Synthesis and characterization of carboxymethyl chitosan carrying ricinoleic functions as an emulsifier for azadirachtin. Carbohydr Polym 88:576-582

Firdhouse MJ, Lalitha P (2013) Biosynthesis of silver nanoparticles using the extract of *Alternanthera sessilis*-antiproliferative effect against prostate cancer cells. Cancer nanotechnol 4:137-143

Fouad H, Hongjie L, Hosni D, Wei J, Abbas G, Ga'al H, Jianchu M (2018) Controlling Aedes albopictus and *Culex pipiens pallens* using silver nanoparticles synthesized from aqueous extract of *Cassia fistula* fruit pulp and its mode of action. Artif Cells Nanomed Biotechnol 46:558-567

Fouad HA,Khalil HM, El-GepalyH, Fouad OA (2016) Nanosilica and jasmonic acid as alternative methods for control *Tutaabsoluta*(Meyrick) in tomato crop under field conditions. Arch Phytopathol Plant Prot 49(13-14):362-370

Fröhlich E, Kueznik T, Samberger C, Roblegg E,Wrighton C, Pieber TR (2010) Size-dependent effects of nanoparticles on the activity of cytochrome P450 isoenzymes. Toxicol Applied Pharmacol 242: 326-332

Ga'al H, Fouad H, Tian J, Hu Y, Abbas G, Mo J (2018) Synthesis, characterization and efficacy of silver nanoparticles against *Aedes albopictus* larvae and pupae. Pestic Biochem Physiol 144:49-56

Gahukar RT, Das RK (2020) Plant-derived nanopesticides for agricultural pest control: challenges and prospects. Nanotechnol Environ Eng 5(1):1-9

Ganesan RM, Gurumallesh Prabu H (2015) Synthesis of gold nanoparticles using herbal *Acorus calamus* rhizome extract and coating on cotton fabric for antibacterial and UV blocking applications. Arabian J Chem 12:2166-2174

Gawade VV, Gavade NL, Shinde HM, Babar SB, Kadam AN, Garadkar KM (2017) Green synthesis of ZnO nanoparticles by using *Calotropis procera* leaves for the photodegradation of methyl orange. J Mater Sci: Mater Electron 28:14033-14039

Ghosh D, Mondal AK (2015) Biosynthesis of silver nano-particle from four medicinally important taxa in South West Bengal, India. J Nanosci Nanotechnol 5:305-312

Gunaselvi G, Kulasekaren V, Gopal V (2010) Anti-bacterial and antifungal activity of various leaves extracts of *Hardwickia binata Roxb.* (Caesalpinaceae). International J Pharm Tech Res 2:2183-2187

Hafez IH, Osman AR, Sewedan EA, Berber MR(2018) Tailoring of a potential nanoformulated form of gibberellic acid: synthesis, characterization, and field applications on vegetation and flowering. J Agric Food Chem 66(31):8237-8245.

Haritha E, Roopan SM, Madhavi G, Elango G, Al-Dhabi NA, Arasu MV (2016) Green chemical approach towards the synthesis of SnO2 NPs in argument with photocatalytic degradation of diazo dye and its kinetic studies. J Photochem Photobiol B Biol 162:441-447

Haryani Y, Nabella I, Yuharmen Y, Kartika GF (2018) Synthesis of silver nanoparticles from *Cucumis melo L.* and assessment of its antimicrobial properties. Pharma Clinical Pharm Res 3:46-50

Hersanti, Djaya L, Hidayat Y, Pratama LS, Joni IM (2020) The effectiveness of suspension of *Beauveria bassiana* mixed with silica nanoparticles (NPs.) and carbon fiber in controlling *Spodoptera litura*. In: AIP Conference Proceedings. AIP Publishing LLC, 2219(1):080011.

Huang J, Li Q, Sun D, Lu Y, Su Y, Yang X, Wang H, Wang Y, Shao W, He N, Hong J, Chen C (2007) Biosynthesis of silver and gold nanoparticles by novel sundried *Cinnamomum camphora* leaf. Nanotechnol 18

Ioset JR, Marston A, Gupta MP, Hostettmann K (2000) Antifungal and larvicidal compounds from the root bark of *Cordia alliodora.* J Nat Prod 63:424–426

Isa AI, Awouafack MD, Dzoyem JP, Aliyu M, Magaji RA, Ayo JO, Eloff JN (2014) Some *Strychnos spinosa* (Loganiaceae) leaf extracts and fractions have good antimicrobial activities and low cytotoxicities. BMC Complement Altern Med 14:456

Ishwarya R, Vaseeharan B, Kalyani S, Banumathi B, Govindarajan M, Alharbi NS, Kadaikunnan S, Al-anbr MN, Khaled JM, Benelli G (2018) Facile green synthesis of zinc oxide nanoparticles using *Ulva lactuca* seaweed extract and evaluation of their photocatalytic, antibiofilm and insecticidal activity. J Photochem Photobiol B Biol 178:249-258

Jayakumarai G, Gokulpriya C, Sudhapriya R, Sharmila G, Muthukumaran C (2015) Phytofabrication and characterization of monodisperse copper oxide nanoparticles using Albizia lebbeck leaf extract. Appl Nanosci 5:1017-1021

Jenne M, Kambham M, Tollamadugu NP, Karanam HP, Tirupati MK, Balam RR, Shameer S,Yagireddy M (2018) The use of slow releasing nanoparticle encapsulated Azadirachtin formulations for themanagement of*Caryedon serratus*O.(groundnut bruchid). IET Nanobiotechnol 12:963-967

Jia X, Sheng WB, Li W, Tong YB, Liu ZY, Zhou F (2014) Adhesive polydopamine coated avermectinmicrocapsules for prolonging foliar pesticide retention. ACS Appl Mater Interfaces 6:19552-19558

Joshi N, Jain N, Pathak A, Singh J, Prasad R, Upadhyaya CP (2018) Biosynthesis of silver nanoparticles using *Carissa carandas* berries and its potential antibacterial activities. J Sol-Gel Sci Tech 86:682-689

Kalimuthu K, Panneerselvam C, Chou C, Tseng LC, Murugan K, Tsai KH, Alarfaj AA, Higuchi A, Canale A, Hwang JS, Benelli G (2017) Control of dengue and Zika virus vector Aedes aegypti using the predatory copepod *Megacyclops formosanus*: synergy with *Hedychium coronarium*-synthesized silver nanoparticles and related histological changes in targeted mosquitoes. Proc Saf Environ Prot 109:82-96

Kamarudin N, Jusoh R, Setiabudi H, Jusoh N, Jaafar N, Sukor N (2019)*Cymbopogon nardus* mediated synthesis of Ag nanoparticles for the photocatalytic degradation of 2,4-dicholorophenoxyacetic acid. Bull Chem Reaction Eng Catal 14:173-181

Kamau RW, Juma BF, Baraza LD (2016) Antimicrobial compounds from root, stem bark and seeds of *Melia volkensii.* Nat Prod Res 30:1984-1987

Kannan BN, Thoppil J (2018) Plant-mediated synthesis of silver nanoparticles by two species of *Cynanchum* L. (Apocynaceae): A comparative approach on its physical characteristics. Int J Nano Dimens 9:104-111

Kantrao S, Ravindra MA, Akbar SMD, Jayanthi PK, Venkataraman A (2017) Effect of biosynthesized Silver nanoparticles on growth and development of *Helicoverpa armigera* (Lepidoptera: Noctuidae): Interaction with midgut protease. J Asia-Pacific Entomol 20(2):583-589

Kar PK, Murmu S, Saha S, Tandon V, Acharya K (2014) Anthelmintic Efficacy of Gold Nanoparticles Derived from a Phytopathogenic Fungus, *Nigrospora oryzae*. PLoS One9

Karthikeyan A, Siva G, Sujitha S, Rex D, Chidambaranathan N (2008) Antihyperglycemic activity of the ethanolic seed extract of *Vernonia anthelminticum* willd. Int J Green Pharm 2:209

Kasithevar M, Saravanan M, Prakash P, Kumar H, Ovais M, Barabadi H, Shinwari ZK (2017) Green synthesis of silver nanoparticles using *Alysicarpus monilifer* leaf extract and its antibacterial activity against MRSA and CoNS isolates in HIV patients. J Interdisciplinary Nanomed: 31-141

Kataria B, Shyam V, Kaushik B, Vasoya J, Joseph J, Savaliya C, Kumar S, Parikh S, Thakar C, Pandya D, Ravalia A, Markna J, Shah N (2017) Green synthesis of silver nanoparticle using *Bambusa arundinacea* leaves. In: Proceedings of the International Conference on Functional Oxides and Nanomaterials, 1837(1): 040043, AIP Publishing LLC.

Kaziem AE, GaoY, Zhang Y, Qin X, Xiao Y, Zhang Y, You H, Li J, He S (2018) Amylase triggeredcarriers based on cyclodextrin anchored hollow mesoporous silica for enhancing insecticidal activity ofavermectin against *Plutella xylostella*. J Hazard Mater 359:213-221

Khan S, Ismail M, Khan T, Hussain F, Majid A, Iqbal T, Tasleem F, Wahab Z, Fayaz M, Rahim T, Khan Y (2018) Green synthesis of Brassica campestris mediated silver nanoparticles, their antibacterial and antioxidant activities. Int J Chem Studies 6(4):1943-1949

Khan SU, Anjum SI, Ansari MJ, Khan MH, Kamal S, Rahman K, Shoaib M, Man S, Khan AJ, Khan SU, Khan D (2019) Antimicrobial potentials of medicinal plant's extract and their derived silver nanoparticles: A focus on honey bee pathogen. Saudi J Biol Sci 26:1815-1834

Kitherian S (2016) Nano and bio-nanoparticles for insect control.Res J Nanosci Nanotech 7:1-9

Kora AJ, Sashidhar R, Arunachalam J (2012) Aqueous extract of gum olibanum (*Boswellia serrata*): A reductant and stabilizer for the biosynthesis of antibacterial silver nanoparticles. Process Biochem 47:1516-1520

Krishnamurthy S, Veerasamy M, Karruppaya G (2020) A Review on Plant Sources for Nano Biopesticide Production. Lett Appl Nano Bio Sci 9:1348-1358

Kulkarni RR, Pawar PV, Joseph MP, Akulwad AK, Sen A, Joshi SP (2013) *Lavandula gibsoni* and *Plectranthus mollis* essential oils: chemical analysis and insect control activities against *Aedes aegypti*, *Anopheles sfttephensi* and *Culex quinquefasciatus*. J Pest Sci 86:713-718

Kumar G, Badoni PP (2018) *Arisaema tortuosum* leaf extract mediated synthesis of silver nanoparticles, characterization and their antibacterial activity. Asian J Res Chem11:419-422

Kumar S, Singh M, Halder D, Mitra A (2016)*Lippia javanica*: a cheap natural source for the synthesis of antibacterial silver nanocolloid. Appl Nanosci 6:1001-1007

Kumar VA, Ammani K, Jobina R, Subhaswaraj P, Siddhardha B (2017) Photo-induced and phytomediated synthesis of silver nanoparticles using *Derris trifoliata* leaf extract and its larvicidal activity against Aedes aegypti. J Photochem Photobiol Bio 171:1-8

Kyomuhimbo HD, Michira IN, Mwaura FB, Derese S, Feleni U, Iwuoha EI (2019) Silver-zinc oxide nanocomposite antiseptic from the extract of Bidens pilosa. SN Appl Sci1

Li F, Gu Z, Wang B, Xie Y, Ma L, Xu K, Ni M, Zhang H, Shen W, Li B (2014) Effects of the biosynthesis and signaling pathway of ecdysterone on silkworm (*Bombyx mori*) following exposure to titanium dioxide nanoparticles. J Chem Ecol 40:913-922

Liang J, Yu M, Guo L, Cui B, Zhao X, Sun C, Wang Y, Liu G, Cui H, Zeng Z (2018) Bioinspireddevelopment of P (St–MAA)–avermectin nanoparticles with high affinity for foliage to enhance foliaretention. J Agric Food Chem 66:6578-6584

Lingaraju K, Raja Naika H, Manjunath K, Basavaraj RB, Nagabhushana H, Nagaraju G, Suresh D (2016) Biogenic synthesis of zinc oxide nanoparticles using *Ruta graveolens* (L.) and their antibacterial and antioxidant activities. Appl Nanosci 6:03-710

Liu X, Vinson D, Abt D, Hurt RH, Rand DM (2009) Differential toxicity of carbon nanomaterials in Drosophila: larval dietary uptake is benign, but adult exposure causes locomotor impairment and mortality. Environ Sci Technol 43:6357-6363

Lu W, LuML, Zhang QP, Tian YQ, Zhang ZX, Xu HH (2013) Octahydrogenated retinoic acid-conjugatedglycol chitosan nanoparticles as a novel carrier of azadirachtin: Synthesis, characterization, and in vitroevaluation. J Polym Sci Part A Polym Chem 51:3932-3940

Mahamadi C, Wunganayi T (2018) Green synthesis of silver nanoparticles using *Zanthoxylum chalybeum* and their antiprolytic and antibiotic properties. Cogent Chem 4

Malleshappa J, Nagabhushana H, Sharma S, Vidya Y, Anantharaju K, Prashantha S, Prasad BD, Naika HR, Lingaraju K, Surendra B (2015) *Leucas aspera* mediated multifunctional CeO_2 nanoparticles: Structural, photoluminescent, photocatalytic and antibacterial properties. Spectrochim Acta Part A Mol Biomol Spectrosc149:452-462

Mao BH, Chen ZY, Wang YJ, Yan SJ (2018) Silver nanoparticles have lethal and sublethal adverse effects on development and longevity by inducing ROS-mediated stress responses. Sci Rep 8(1):2445

Marimuthu S, Rahuman AA, Rajakumar G, Santhoshkumar T, Kirthi AV, Jayaseelan C, Bagavan A, Zahir AA, Elango G, Kamaraj C (2011) Evaluation of green synthesized silver nanoparticles against parasites. Parasitol Res 108(6):1541-1549

Mohan SC, Hemalatha K, Manivel A, Kumar MS (2018) Synthesis and characterization of Sn/ZnO nanoparticles for removal of organic dye and heavy metal. Int J Biol Chem 12:1-7

Mommaerts V, Jodko K, Thomassen LC,Martens JA, Kirsch-Volders M, SmaggheG (2012) Assessment of side-effects by LudoxTMA silica nanoparticles following a dietary exposure on the bumblebee *Bombus terrestris*. Nanotoxicol 6:554-561

Naghdi S, Sajjadi M, Nasrollahzadeh M, Rhee KY, Sajadi SM, Jaleh B (2018) *Cuscuta reflexa* leaf extract mediated green synthesis of the Cu nanoparticles on graphene oxide/manganese dioxide nanocomposite and its catalytic activity toward reduction of nitroarenes and organic dyes. J Taiwan Inst Chem Eng 86:158-173

Nair PMG, Choi J (2011) Identification, characterization and expression profiles of *Chironomus riparius* glutathione S-transferase (GST) genes in response to cadmium and silver nanoparticles exposure. Aquat Toxicol 101(3):550-560

Nair PMG, Choi J (2012) Modulation in the mRNA expression of ecdysone receptor gene in aquatic midge, *Chironomus riparius* upon exposure to nonylphenol and silver nanoparticles. Environ Toxicol Pharmacol 33:98-106

Nair PMG, Park SY, Choi J (2013) Evaluation of the effect of silver nanoparticles and silver ions using stress responsive gene expression in *Chironomus riparius*. Chemosphere 92:592-599

Nair PMG, Park SY, Lee SW, Choi J (2011) Differential expression of ribosomal protein gene, gonadotrophin releasing hormone gene and Balbiani ring protein gene in silver nanoparticles exposed *Chironomus riparius*. Aquat Toxicol 101:31-37

Nair RS, Snima KS, Kamath RC, Nair SV, Lakshmanan VK (2015) Synthesis and characterization of *Careya Arborea* nanoparticles for assessing its in vitro efficacy in Pancreatic Cancer Cells. J Nat Prod 8

Nasrollahzadeh M, Issaabadi Z, Sajadi SM (2018) Green synthesis of a Cu/MgO nanocomposite by *Cassytha filiformis* L. extract and investigation of its catalytic activity in the reduction of methylene blue, congo red and nitro compounds in aqueous media. RSC Adv 8:3723-3735

Nasrollahzadeh M, Sajadi SM (2016) Green synthesis of Pd nanoparticles mediated by *Euphorbia thymifolia* L. leaf extract: Catalytic activity for cyanation of aryl iodides under ligand-free conditions. J Colloid Inter Sci 469:191-195

Nguyen TNQ, Le VA, Hua QC, Nguyen TT (2015) Enhancing Insecticide Activity of Anacardic Acid byIntercalating it into MgAl Layered Double Hydroxides Nanoparticles; Institut für Abfallwirtschaft und Altlasten, Fakultät Umweltwissenschaften, Technische Universität Dresden: Dresden, Germany, 2015.

Ojewole JAO (2008) Analgesic, anti-inflammatory and hypoglycaemic effects of *Securidaca longepedunculata* (Fresen.) [Polygalaceae] root-bark aqueous extract. InflammoPharmacol16:174-181

Ojha S, Sett A, Bora U (2017) Green synthesis of silver nanoparticles by *Ricinus communis* var. carmencita leaf extract and its antibacterial study. Adv Nat Sci: Nanosci Nanotech 8

Oliveira JLD, Campos ENV, Pereira AE, Pasquoto T, Lima R, Grillo R, Andrade DJD, Santos FAD, Fraceto LF (2018) Zein nanoparticles as eco-friendly carrier systems for botanical repellentsaiming sustainable agriculture. J Agric Food Chem 66:1330-1340

Palermo D, Giunti G, Laudani F, Palmeri V, Campolo O (2021) Essential oil-based nano-biopesticides: Formulation and bioactivity against the confused flour beetle *Tribolium confusum*. Sustainability13(17):9746

Panda SK, Dutta SK, Bastia AK (2010) Antibacterial activity of *Croton roxburghii* Balak. against the enteric pathogens. J Adv Pharma Tech Res1:419-422

Panda SK, Mohanta YK, Padhi L, Park YH, Mohanta TK, Bae H (2016) Large scale screening of ethnomedicinal plants for identification of potential antibacterial compounds. Molecules 21

Parthiban E, Manivannan N, Ramanibai R, Mathivanan N (2019) Green synthesis of silver-nanoparticles from *Annona reticulata* leaves aqueous extract and its mosquito larvicidal and anti-microbial activity on human pathogens. Biotechnol Rep 21

Patel MH, Vashi KD, Minocheherhomji FP, Nagnae R (2019) Evaluation of antimicrobial activity and mutagenicity of copper nanoparticle synthesized using green chemistry. Int J Recent Sci Res 10:31196-31200

Patil CD, Borase HP, Suryawanshi RK, Patil SV (2016) Trypsin inactivation by latex fabricated gold nanoparticles: a new strategy towards insect control. Enzym Microb Technol 92:18-25

Paul M, Londhe VY (2018) *Pongamia pinnata* seed extract-mediated green synthesis of silver nanoparticles: Preparation, formulation and evaluation of bactericidal and wound healing potential. ApplOrganometallic Chem33

Paulraj MG, Ignacimuthu S, Gandhi MR, Shajahan A, Ganesan P, Packiam SM, Al-Dhabi NA (2017)Comparative studies of Tripolyphosphate and Glutaraldehyde cross-linked chitosan-botanical pesticide nanoparticles and their agricultural applications. Int J Boil Macromol 104(B):1813-1819

Philbrook NA,Winn LM, Afrooz AN, Saleh NB,Walker VK (2011) The effect of TiO2 and Ag nanoparticles on reproduction and development of *Drosophila melanogaster* and CD-1 mice. Toxicol Appl Pharmacol 257:429-436

Prabhu D, Arulvasu C, Babu G, Manikandan R, Srinivasan P (2013) Biologically synthesized green silver nanoparticles from leaf extract of *Vitex negundo* L. induce growth-inhibitory effect on human colon cancer cell line HCT15. Process Bioche 48:317-324

Rajkuberan C, Prabukumar S, Sathishkumar G, Wilson A, Ravindran K, Sivaramakrishnan S (2017) Facile synthesis of silver nanoparticles using *Euphorbia antiquorum* L. latex extract and evaluation of their biomedical perspectives as anticancer agents. J Saudi Chem Soc 21:911-919

Rani PU, Madhusudhanamurthy J, Sreedhar B (2014) Dynamic adsorption of alpha-pinene and linalool on silicananoparticles for enhanced antifeedant activity against agricultural pests. J Pest Sci 87:191-200

Rathnasamy R, Thangasamy P, Thangamuthu R, Sampath S, Alagan V (2017) Green synthesis of ZnO nanoparticles using *Carica papaya* leaf extracts for photocatalytic and photovoltaic applications. J Master Sci: Master Electron 28:10374-10381

Ravindran CP, Manokari M, Mahipal S, Shekhawat (2016) Biogenic production of zinc oxide nanoparticles from aqueous extracts of *Duranta erecta* L. World Sci News 28:30-40

Sarkar D (2017) Synthesis of plant-mediated silver nanoparticles using *Commiphora Wightii* (Guggul) extract and study their antibacterial activities against few selected organisms. World J Pharma Pharmac Sci 1418-1425

SelvaraniS, Vinayaga PM, RamalakshmiM, Padmapriya S, Abirami M (2016)*Ocimum Kilimandscharicum* leaf extract engineered silver nanoparticles and its bioactivity. J Nanomater Mol Nanotechnol 5

ShahzadiI, Shah MM (2015) Acylated flavonol glycosides from *Tagetes minuta* with antibacterial activity. Front Pharma6

Shaikh R, Pund M, Dawane A, Iliyas S (2014) Evaluation of anticancer, antioxidant, and possible anti-inflammatory properties of selected medicinal plants used in Indian traditional medication. J Traditional Complementary Med 4:253-257

Shameli K, Ahmad MB, Zamanian A, Sangpour P, Shabanzadeh P, Abdollahi Y, Zargar M (2012) Green biosynthesis of silver nanoparticles using *Curcuma longa* tuber powder. Int J Nanomed 7:5603-5610

Sharma M, Yadav S, Ganesh N, Srivastava MM, Srivastava S (2019) Biofabrication and characterization of flavonoid-loaded Ag, Au, Au–Ag bimetallic nanoparticles using seed extract of the plant *Madhuca longifolia* for the enhancement in wound healing bio-efficacy. Prog Biomater 8:51-63

Sheng WB, Li W, Zhang GX, Tong YB, Liu ZY, Jia X (2015) Study on the UV-shielding andcontrolled-release properties of a polydopamine coating for avermectin. New J Chem 39:2752-2757

Shoaib A, Elabasy A, Waqas M, LinL, Cheng X, Zhang Q, Shi ZH (2018) Entomotoxic effect of silicon dioxide nanoparticles on *Plutella xylostella* (L.) (Lepidoptera: Plutellidae) under laboratory conditions. Toxicol Environ Chem 100:80-91

Singh H, Du J, Singh P, Yi TH (2018) Role of green silver nanoparticles synthesized from *Symphytum officinale* leaf extract in protection against UVB-induced photoaging. J Nanostruct Chem 8:359-368

Singh P, Pandit S, Garnæs J, Tunjic S, Mokkapati VR, Sultan A, Thygesen A, Mackevica A, Mateiu RV, Daugaard AE, Baun A, Mijakovic I (2018) Green synthesis of gold and silver nanoparticles from *Cannabis sativa* (industrial hemp) and their capacity for biofilm inhibition. Int J Nanomed 13:3571-3591

Singhal G, Bhavesh R, Kasariya K, Sharma AR, Singh RP (2011) Biosynthesis of silver nanoparticles using *Ocimum sanctum* (Tulsi) leaf extract and screening its antimicrobial activity. J Nanopart Res 13:2981-2988

Small T, Ochoa-Zapater MA, Gallello G, Ribera A, Romero FM, Torreblanca A, Garcerá MD (2016) Gold-nanoparticles ingestion disrupts reproduction and development in the German cockroach. Sci Total Environ 565:882-888

Souza M, Lopes L, Bonez P, Gündel A, Martinez D, Sagrillo M, Giongo J, Vaucher R, Raffin R, Boligon A, Santos R (2017) *Melaleuca alternifolia* nanoparticles against Candida species biofilms. Microb Pathogen 104:125-132

Srikar SK, Giri DD, Pal DB, Mishra PK, Upadhyay SN (2016) Green Synthesis of Silver Nanoparticles: A Review. Green Sustainable Chem 6:34-56

Stadler T, Lopez-Garcia GP, Gitto JG, Buteler M (2017) Nanostructured alumina: biocidal properties and mechanism of action of a novel insecticide powder. Bull Insectol 70:17-25

Steffy K, Shanthi G, MarokyAS, Selvakumar S (2018) Synthesis and characterization of ZnO phytonanocomposite using *Strychnos nux-vomica* L. (Loganiaceae) and antimicrobial activity against multidrug-resistant bacterial strains from diabetic foot ulcer. JAdv Res 9:69-77

Sujitha MV, Kannan S (2013) Green synthesis of gold nanoparticles using Citrus fruits (*Citrus limon, Citrus reticulata* and *Citrus sinensis*) aqueous extract and its characterization. Spectrochim Acta Part A Mol Biomol Spectrosc 102:15-23

Sultana N, Raul PK, Goswami D, Das B, Gogoi HK, Raju PS (2018) Nanoweapon: control of mosquito breeding using carbon-dot-silver nanohybrid as a biolarvicide. Environ Chem Lett. 16(3):1017-1023.

Sundararajan B, Kumari BR (2017) Novel synthesis of gold nanoparticles using *Artemisia vulgaris* L. leaf extract and their efficacy of larvicidal activity against dengue fever vector *Aedes aegypti* L. J Trace Elem Med Biol 43:187-196

Suresh U, Murugan K, Panneerselvam C, Rajaganesh R, Roni M, Al-Aoh HAN, Trivedi S, Rehman H, Kumar S, Higuchi A (2018) Suaeda maritima-based herbal coils and green nanoparticles as potential biopesticides against the dengue vector *Aedes aegypti* and the tobacco cutworm *Spodoptera litura*. Physiol Mol Plant Pathol 101:225-235

Tippayawat P, Phromviyo N, Boueroy P, Chompoosor A (2016) Green synthesis of silver nanoparticles in *aloe vera* plant extract prepared by a hydrothermal method and their synergistic antibacterial activity. PeerJ 4:e2589

Tran TTT, Vu TTH, Nguyen TH (2013) Biosynthesis of silver nanoparticles using *Tithonia diversifolia* leaf extract and their antimicrobial activity. Mater Lett 105:220-223

Ulrichs C, Mewis I, Goswami A (2005) Crop diversification aiming nutritional security in West Bengal-biotechnology of stinging capsules in nature's water-blooms. Ann Tech Issue of State Agri Technologists Service Assoc1-18

Umar H, Kavaz D, Rizaner N (2018) Biosynthesis of zinc oxide nanoparticles using *Albizia lebbeck* stem bark, and evaluation of its antimicrobial, antioxidant, and cytotoxic activities on human breast cancer cell lines. Int J Nanomed 14:87-100

Venkat Kumar S, Karpagambigai S, Jacquline Rosy P, RajeshkumarS (2017) Phyto-assisted synthesis of silver nanoparticles using *Solanum Nigrum* and antibacterial activity against *Salmonella Typhi* and Staphylococcus Aureus. MechMater Sci Eng J 9

Venkata Subbaiah KP, Savithramma N (2013) Characterization and validation of silver nanoparticles from *holarrhena pubescens* – an important ethnomedicinal plant of Kurnool district, AndhraPradesh, India. World J Pharma Pharmac Sci 2:6288-6300

Vijayakumar M, Priya K, Nancy F, Noorlidah A, Ahmed A (2013) Biosynthesis, characterisation and anti-bacterial effect of plant-mediated silver nanoparticles using *Artemisia nilagirica*. Ind Crops Prod 41:235-240

Vijayakumari A, Sinthiya A (2018) Biosynthesis of Phytochemicals Coated Silver Nanoparticles Using Aqueous Extract of Leaves of *Cassia alata* – Characterization, Antibacterial and Antioxidant Activities. Int JPharmac Clinical Res10:138-149

Vinayaga Moorthi P, Balasubramaniam C, Mohan S (2015) An improved insecticidal activity of silver nanoparticle synthesized by using *Sargassum muticum*. Appl Biochem Biotechnol 175:135-140

Vivek R, Thangam R, Muthuchelian K, Gunasekaran P, Kaveri K, Kannan S (2012) Green biosynthesis of silver nanoparticles from *Annona squamosa* leaf extract and its in vitro cytotoxic effect on MCF-7 cells. Process Biochem 47:2405-2410

Wang Y, Cui H, Sun C, Zhao X, Cui B (2014) Construction and evaluation of controlled-release delivery systemof Abamectin using porous silica nanoparticles as carriers. Nanoscale Res Lett 9:2490

Wardani M, Yulizar Y, Abdullah I, Bagus Apriandanu D (2019) Synthesis of NiO nanoparticles via green route using *Ageratum conyzoides* L. leaf extract and their catalytic activity. IOP Conference Series: Mater Sci Eng 509

Widyaningtyas AL, Yulizar Y, Apriandanu DOB (2019) Ag_2O nanoparticles fabrication by *Vernonia amygdalina* Del. leaf extract: synthesis, characterization, and its photocatalytic activities. IOP Conference Series: Mater Sci Eng 509

Worrall EA, Hamid A, Mody KT, Mitter N, Pappu HR (2018) Nanotechnology for plant disease management. Agron 8:285

Yang FL, Li XG, Zhu F, Lei CL (2009) Structural characterization of nanoparticles loaded with garlicessential oil and their insecticidal activity against *Tribolium castaneum* (Herbst) (Coleoptera: Tenebrionidae). J Agric Food Chem 57:10156-10162

Yang Y, Cheng J, Garamus VM, Li N, Zou A (2018) Preparation of an environmentally friendly formulation ofthe insecticide nicotine hydrochloride through encapsulation in chitosan/ tripolyphosphate nanoparticles. J Agric Food Chem 66:1067-1074

Yasur J, Usha-Rani P (2015) Lepidopteran insect susceptibility to silver nanoparticles and measurement of changes in their growth, development and physiology. Chemosphere 124:92-102.

Yu C, Tang J, Liu X, Ren X, Zhen M, Wang L (2019) Green biosynthesis of silver nanoparticles using *Eriobotrya japonica* (Thunb.) leaf extract for reductive catalysis. Mater 12

Yuan CG, Huo C, Yu S, Gui B (2017) Biosynthesis of gold nanoparticles using *Capsicum annuum* var. grossum pulp extract and its catalytic activity. Physica E: Low-dimensional Systems Nanostruct 85:19–26.

Zaki AM, Zaki AH, Farghali AA, Abdel-Rahim EF (2017) Sodium titanate-bacillus as a new nanopesticide for cotton leaf-worm. J Pure Appl Microbiol11(2):725-32.

ZhangH, Qin H, Li L, Zhou X, Wang W, Kan C (2018) Preparation and characterization of controlled-releaseavermectin/castor oil-based polyurethane nanoemulsions. J AgricFood Chem 66:6552-6560.

Zhao L, Peralta-Videa JR, Rico CM, Hern andez-Viezcas JA, Sun Y, Niu G, Servin A, Nunez JE, Duarte-Gardea M, Gardea-Torresdey JL (2014) CeO_2 and ZnO nanoparticles change the nutritional qualities of cucumber (*Cucumis sativus*). J AgricFood Chem 62:2752-2759.

Zhu H, Shen Y, Cui J, Wang A, Li N, Wang C, Cui B, Sun C, Zhao X, Wang C, Gao F, Zhan S, Guo L, Zhang L, Zeng Z, Wang Y, Cui H (2020) Avermectin loaded carboxymethyl cellulose nanoparticles with stimuli-responsive and controlled release properties. Ind Crop Prod 152:112497.

Zinjarde SS, Bhargava SY, Kumar AR (2011) Potent α-amylase inhibitory activity of Indian Ayurvedic medicinal plants. BMC Complement Altern Med 11.

6

Impact of Biopesticides Application on Crop Quality and Environmental Quality

Abstract

Synthetic organic insecticides are posing serious health hazards to human beings and other non-target organisms across the world. The consumers are well aware of the ill effects of insecticides and there is a great demand for pesticide free agricultural produces. In few occasions, Indian consignments such as chillies, mango, vegetables, wheat have been either rejected by the importing countries or in the international markets due to pesticides residues, quarantine pests etc. Insecticides of biological origin such biopesticides are gaining importance as they are environmentally safe, leaving no residues in products etc. To date more than 12 biopesticides have been registered in India for their use in pest management. Many more are in pipeline to combat the yield losses caused by pests. Some of the biopesticides such EPN's have controlled white grubs without compromising soil and environmental quality. Biopesticides such as Bacillus subtilis GA1 and Bacillus sp. have well preserved the mango juice over a period of 15 days which was as good as chemical preservative there by improved the shelf life of perishable fruits. The government of India has given greater emphasis on natural farming/zero budget farming, organic farming where biopesticides play crucial role in management of pest populations. In this chapter, we have narrated the economic importance of biopesticides, international trade issues, impact of biopesticides on crop quality, soil and environmental quality.

Keywords: Biopesticides, Soil, Environment, Quality, Contamination

Introduction

Insecticide based plant protection in India as well as in the world resulted in wide variety of environmental and health issues. Insecticides although gave satisfactory control of target pests initially, have posed serious issues such as resistance, resurgence and residue and also contamination of water bodies, food chain leading serious ill effects in humans, mammals, soil beneficial microbiota etc. After thorough examination of scientific evidences and facts on ill effects of synthetic organic insecticides, much emphasis was given for biological control including parasitoids, predators, entomopathogens, biopesticides, microbial control etc. in order to overcome above ill effects. The concept of biopesticides have come up as an alternative to the indiscriminate use of harmful synthetic insecticides which are being extensively used in organic and natural/zero budget farming. Biopesticides are formulations made from naturally occurring substances like animals, plants, microorganisms and include living organisms, their products or byproducts that control pests by non-toxic mechanisms in an ecofriendly manner. The biopesticides such as botanicals, Entomopathogenic fungi, *Bt*, NPV, EPN, PI, etc. which are widely used in the global market including India. Biopesticides may be categorized into three major groups: plant-incorporated protectants (PIPs), biochemical, and microbial biopesticides. While microbial biopesticides use microorganisms (bacteria, fungi, viruses or protozoans) as active-ingredient, biochemical pesticides are naturally occurring substances from plants and animals. PIPs are produced naturally on genetic modification of a crop plant, such as *Bt* cotton. Such transgenic plant produces biodegradable protein with no harmful effect on animals and human beings, and thus curtails the use of hazardous pesticides. PIPs may be more effective and economical strategies in the developing countries to help produce more food, feed and forages in an environmentally safer manner. However, the pesticides of biological origin have also been reported to cause relatively less side effects which are being discussed in this chapter.

Merits of Biopesticides

The biopesticides are more preferred in today's health conscious world due to following advantages.

- Biopesticides are generally less toxic than chemical pesticides often target specific pests
- Little or no residual effects hence pose less risks to human health and environment
- Have wide acceptability for use in the organic farming
- Many biopesticides have a zero or low re-entry and handling interval
- Some microbial biopesticides can reproduce on or near to the target pest / disease, giving some self-perpetuating control

- The risk of pests and disease developing resistance to biopesticides is often considered to be low
- They often have good compatibility both with biological pest control agents (natural enemies) and conventional chemical pesticides, so can be readily incorporated into IPM
- They can also be useful as a second line of defence or supplementary treatment
- Relatively less costly
- Enhanced crop quality
- Adequately degradable
- No harmful residues remain in food, fodder and fibers
- Growing market preferences

Constraints in Exporting Agricultural Commodities

India is one of the largest producers of a number of agriculture commodities and the EuropeanUnion (EU) is one of the largest export markets for India. India is seeing growth in the export of agricultural commodities like cereals, non-basmati rice, wheat, millets, maize, and other coarse grains and the largest markets for India's agricultural products are the US, China, Bangladesh, the UAE, Vietnam, Saudi Arabia, Indonesia, Nepal, Iran, and Malaysia.Demand for Indian cereals was robust in 2020-21, with shipments sent to several countries for the first time, such as rice to countries like Timor-Leste, Puerto Rico, and Brazil. Similarly, wheat was despatched to countries such as Yemen, Indonesia, and Bhutan, and other cereals have been exported to Sudan, Poland, Bolivia. In financial year 2021, fresh fruits were the leading horticulture product exported from India (56 billion Indian rupees). Over 956 thousand metric tons of fruits were exported that year from the south Asian country. Organic exports that include products such as cereals and millets, spices and condiments, tea, medicinal plant products, dry fruits, and sugar grew 51 per cent year on year to $1,040 million. However, the pesticide residues in the commodities are the major bottlenecks in exporting the quality products to international market. Pesticide residue problems have affected exports of basmati rice which is the key traditional export product to the EU, due to stringent norms imposed for chemicals such as Tricyclazole and Buprofezin, extensively used in rice cultivation in India. Testing by the Export Inspection Council (EIC) has been made mandatory for basmati exports to the EU, which led to a decrease in the number of alerts.

In recent years a number of Indian agricultural products have been facing rejection and export bans in the EU due to standards related to food quality, safety and health, sanitary and phytosanitary (SPS) measures. The products such as mangoes, grapes and eggplants in which Indian exporters have faced rejections or bans in the EU and other markets in the past for SPS issues such as fruit flies or thrips infestation. Among the pesticides/chemicals, aflatoxins had the maximum notifications in Basmati rice, followed by Carbendazim, Acephate, Triazophos, Hexaconazole and other miscellaneous pesticides (such as bromide, chlorpyrifos, ochratoxin and profenofos).. EU rejected table grapes consignments from India in 2010, leading to a slowdown in the industry. Reduction in the chlormequat chloride limits in grapes from 0.05mg/kg to0.01mg/kg in the year 2016 which hampered the export of grapes.Saudi Arabia actually cited pesticide levels beyond its own MRLs to block shipments of green chillies and cardamom from India. In the EU, on the other hand, Indian export products that have faced issues on MRL levels in 2020 include:

- Sesame Seeds: Ethylene Oxide (insecticide)
- Chillies: Chlorothalonil (fungicide)
- Frozen curry leaves: Chlorpyrifos (pesticide)
- Frozen diced red chilli puree: Methamidophos, monocrotophos, acephate, propargite and triazophos
- Basmati Rice: Thiamethoxam, tricyclazole and buprofezin

Impact of Biopesticides on Crop Quality

Crop quality is of utmost importance to both growers and consumers. Plant physiology is highly responsive to the prevailing environmental conditions that plays a critical role in both quantity and quality. Active management of plant physiology plays an important role in crop productivity, and biopesticides, particularly those in the plant growth regulator category (PGRs), are key tools in this regard. Characteristics such as fruit size, taste, texture, shape, colour, firmness and shelf life can all be enhanced by careful use of plant growth regulators. In addition, some PGRs can give a boost to plant health by increasing the root mass or enhancing resistance to pests and disease. PGRs have the added benefit of being non-toxic. No harmful residues remain to delay handling or consumption. Crop quality and yield largely determine a grower's income. Biopesticides provide dealers with products that can markedly improve crop quality and yield by preventing pest damage and promoting physiological benefits in plants, including increased fruit size and enhanced colour. Dealers who supply biopesticides and encourage their innovative use are on the forefront of yield-and-profit enhancement practices. Most bio-based pest management products are listed for use in organic farming, providing those growers with compelling pest control options to protect yields and quality. A heterogeneous representation of target products, such as winter guava, mango, apple, mandarin, kiwifruit, strawberry, pepper fruit, red-fleshed table grape, pineapple, cherry fruit, papaya, plum needs much attention in consumer point of view due to ill effects of synthetic pesticides.

Bacillus subtilis GA1 and *Bacillus* sp. have well preserved the mango juice over a period of 15 days which was as good as chemical preservative. Biopesticides serves as good preservatives and attract consumer preferences than chemical (Kohi et al. 2020). Biopesticides such as *T. harzianum* T22 and 6PP are able to improve crop yield and increase the total amount of polyphenols and antioxidant activity in the grapes by reducing the powdery mildew fungi indicting the improvement of crop quality (Pascale et al. 2017). Besides, enhancement of corn yield was reported in several commercial which has been considered as a direct effect of an increased root and foliar systems (Harman 2000). The PGPR activity is induced by *Trichoderma* can be explained by an upregulation of photosynthesis related proteins and a higher photosynthetic efficiency, enhanced the plant nutrient uptake mechanism and increased plant nitrogen use efficiency etc (Harman et al. 2004). Even biopesticides such as foliar spray of neem oil @1.5% along with tree pruning significantly improved fruit physical quality and cosmetic appearance of mandarin (Aftab et al. 2021). Arbuscular mycorrhizal fungi play major role in biological control of plant diseases owing to their capabilities of amelioration crop yields by multiple role as bio-pesticides and plant growth promotion (Nelson 2004). Mycorrhiza can be seen as an assurance against quality deterioration caused by stress factors and also positive effects not only on plant growth, but also on plant quality that include improved product quality of lettuce, tomato, pepper and strawberries (Baum et al. 2015). Preharvest foliar spray of fungal culture filtrates from *Aspergillus niger* and *Rhizopus oryzae* improved the plant defence mechanism, with also enhanced quality and shelf life of date fruit in India (Bhatt and Jampala 2020). Pre-harvest treatment with *Metschnikowia fructicola* for the control of postharvest rots not only reduced the fruit rots significantly but also improved the fruit quality strawberry quality (Karabulut et al. 2004; Sellitto et al. 2021). Botanical and microbial fungicides are effective for suppressing botrytis fruit rot in strawberry alone or in rotation with synthetic fungicides thereby increased the quality for fetching higher market value (Dara 2020).

Impact of Biopesticides on Environmental Quality

Sustainable use of agro-pharmaceuticals, together with the demand for more environment-friendly production systems are the need of the hour in health-conscious consumer world. A growing public interest in the search for alternative approaches to chemical control in biotic stress management is very much required. The time-tested indigenous technical knowledge (ITK) of using natural materials for the control of pests has been very effective which need to be practiced. Biopesticides pose less threat to the environment and human health. They are generally less toxic than chemical pesticides, often target specific, have little or no residual effects and have acceptability for use in organic farming. Use of botanicals is now emerging as one of the important means to be used in protection of crop produce and the environment from pesticidal pollution, which is a global problem. There is less danger of biopesticide impact on the environment and water quality and they offer a more environmentally friendly alternative to chemical insecticide. Biopesticides have long been attracting global attention as a safer strategy than chemical pest control, with potentially less risk to humans and the environment. To this end, co-operation between the public and private sectors is required to facilitate the development, manufacturing, and sale of this environmentally friendly alternative.

Case Studies

Entomopathogenic Nematodes (EPN) for Crop and Soil Health

Entomopathogenic nematodes (EPN) of families Heterorhabditidae and Steinernematidae are microscopic, non-segmented roundworms that are obligate parasites of insects and have become important in biological control and integrated insect pest management as biopesticides. Soil insect pests including white grubs, cutworms, termites, root grubs, etc., cause 24-40% yield losses in sugarcane, maize, arecanut, cardamom, groundnut, potato, banana, guava, turmeric, pulses, vegetables, grasses, lawns etc., and direct plant loss to the tune of 20-60% in arecanut, sugarcane, cardamom, banana, groundnut, turmeric, guava, soybean etc. Due to continuous depletion of forest cover and organic carbon, summarily attributed to anthropogenic and geological events, the soilborne insect pests are increasingly causing a serious threat. Many synthetic chemicals like OP, carbamates, neonicotinoids, fumigants etc., are in indiscriminate use with little effect on the target pest, but causing soil and water body contamination, residual effects on soil biota, human and animal wellbeing, soil health and productivity. Farmers are desperately looking for ecologically safe, sustainable and on-farm recyclable green technologies,

Colour Plates

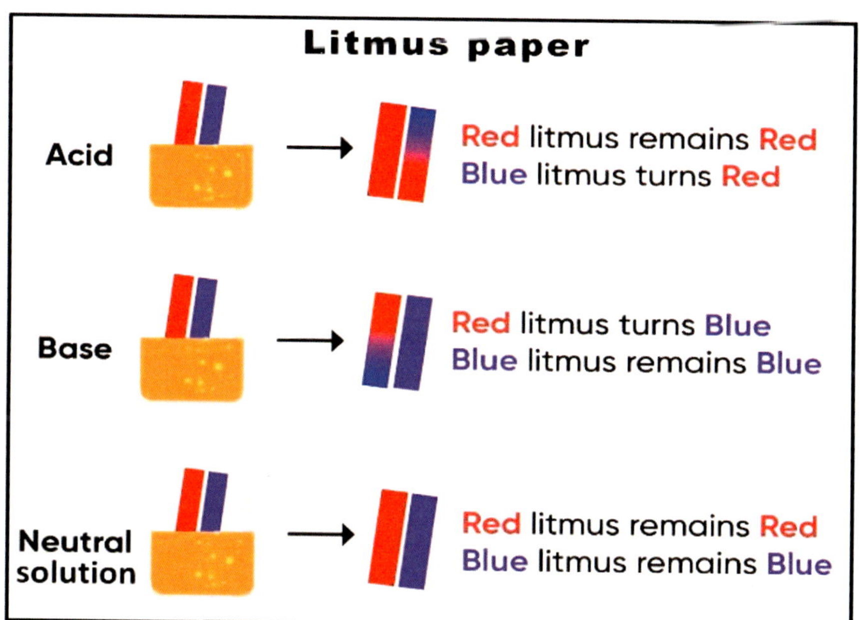

Fig. 10.3: Colour changes on litmus papers

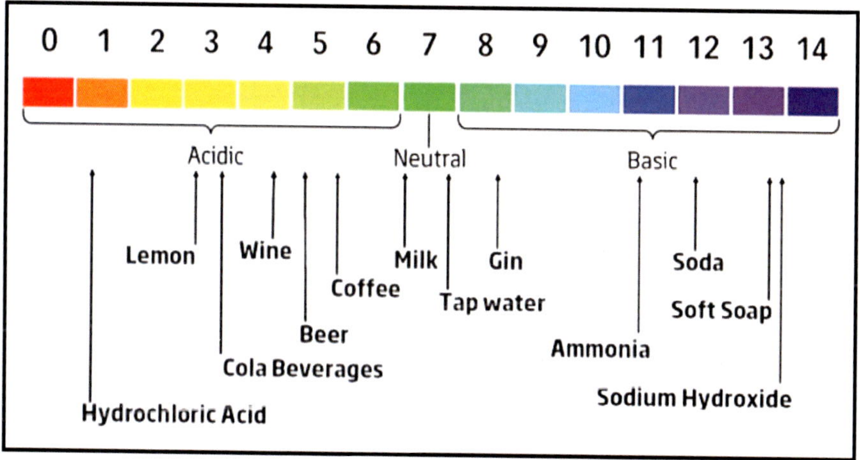

Fig. 10.4: pH of different solutions

7

Indian Biopesticides Market, Consumption, Growth and Opportunities

Abstract

In India, the market for biopesticides has reportedly grown at a quick and rapid rate (23%) over the previous ten years, whereas the market for chemical pesticides has only grown by 2%. However, the usage of biopesticides has not yet reached the same level as that of chemical pesticides, but it is predicted to do so between 2040 and 2050. There is reportedly a 30% difference between the demand for and consumption of microbial biopesticide in India. Anytime the use of biopesticide is encouraged to a broad adoption by stakeholders through appropriate development of awareness, further market sizing is conceivable. Maharashtra state consumed the most biopesticide formulations, whereas Chhattisgarh, Haryana, and other Indian states consumed the least. According to crop consumption, cereal crops receive the most, followed by pulses, oilseeds, fibre, fruits, and vegetables. The market and consumption of biopesticides in India and globally are reported.

Keywords: Biopesticide, Demand and Consumption, State-wise, Crop-wise

Introduction

A tremendous change in the Indian biopesticide industry and market has reflected on increased global trade in agricultural commodities, a healthy revolution in the consumers and stack-holders with adequate understanding on negative effective of chemical pesticides in plant protection. Currently biopesticides comprise approximately 3-5% of the Indian pesticide market, with at least 15 microbial species and 970 microbial formulations registered through the Central Insecticides Board and Registration Committee (CIBRC). As of 2017, over 200 products based on entomopathogenic fungi (*Beauveria bassiana*, *B. brongniartii*, *Metarhizium anisopliae*, *Lecanicillium lecanii* and

Hirsutella thompsonii) and nematicidal fungi (*Purpureocillium lilacinum* and *Pochonia chlamydosporia*) are registered for use against various arthropods and plant parasitic nematodes. Regarding bacteria, over 30 products based on *Bacillus thuringiensis* (Bt) subsp. kurstaki are registered against bollworms, loopers and other Lepidoptera, while 12 based on *Bt* subsp. israelensis and three with *Bt* subsp. sphaericus are being used against mosquitoes.

Two viruses are registered, namely *Helicoverpa armigera* nucleopolyhedron virus (22 products) and *Spodoptera litura* NPV (5 products) for use against bollworms and armyworms. Four entomopathogenic nematode species consisting wettable powder formulations of *Heterorhabditis indica* developed by the ICAR-National Bureau of Agricultural Insect Resources, Bengaluru which have been distributed on a large scale to control white grubs and other sugarcane pests. Biopesticide research in developing countries like India though in infant stage, but evolving rapidly, and focusing on indigenous entomopathogens. Despite enormous regulation, quality-control issues and limited large-scale production facilities, investment in domestic fermentation technologies, improved delivery systems, and promotion of biological control through private and public initiative will increase the share of microbial biopesticides in the country.

Biopesticide Market: Global and Indian Perspectives

Biopesticides are organic substances used to control pests that are derived from plants, animals, microbes, and some minerals. Only one entomopathogenic bacteria, *Bacillus thuringiensis*, is the source of about 90% of the microbial biopesticides currently on the market (Kumar and Singh 2015). Currently, biopesticides only account for a small portion of the overall crop protection business, with a value of roughly $3 billion globally, or 5% of the total crop protection market (Marrone 2014; Olson 2015).

There are more than 200 items accessible on the United States (US) market, compared to only 60 comparable products on the European Union (EU) market. Although the global market for these pesticides appears to need to expand further in the future if these products are to play a significant role in replacing chemical pesticides and lessening the current over-reliance on them, even though the use of biopesticides is rising globally by almost 10% annually (Kumar and Singh 2015). However, it should be noted that the EU uses the same regulations for evaluating biopesticides as they do for synthetic active substances. As a result, several new provisions in the law were needed, and new guidelines were also created to make it easier for potential biopesticide products to be registered (Czaja et al. 2015). The EU currently has less registered biopesticide active chemicals than the US, India, Brazil, or China. The greater complexity of EU-based biopesticide laws is connected to the comparatively low level of biopesticide research in the EU (Balog et al. 2017).

With compounded annual growth rates of more than 15%, biopesticides are expected to outgrow chemical pesticides in terms of growth (Marrone 2014). Between the late 2040s and the early 2050s, it is anticipated that the market size for biopesticides will equal that of synthetics, but there are significant uncertainties surrounding the rates of uptake, particularly in regions like Africa and Southeast Asia, which account for a significant portion of the flexibility in those projections (Olson 2015). In recent years, biopesticides have grown in popularity and are thought to be safer than traditional pesticides. Biopesticides have the potential to reduce the usage of conventional pesticides as essential elements of IPM programmes because they are effective in small amounts and breakdown quickly without leaving harmful residues. However, it should be noted that while there may be situation-specific exceptions to the aforementioned qualities, they do not negate the overall norm.

Indian market is a house to hundreds of biopesticides that are duly registered by the Central Insecticides Board and Registration Committee (CIB&RC), but quality control is a major problem in most of these products. Extensive research on biopesticides in national laboratories and State Agricultural Universities has clearly demonstrated the efficacy of biopesticides for management of pests and diseases. Regardless of the persistent government programs and initiatives, the consumption of biopesticides in India has remained relatively low, for several years in past especially since 2000s. The recent years have witnessed the introduction of nanotechnology mediated biopesticides. Nanoparticles mediated biopesticides have shown considerate potential in alleviating the problems associated with conventional pesticides. The market has attained speedy growth over the period of six years from FY'2013 to FY'2019. The potential benefits of using biopesticides in agriculture and public health programs are considerable. This has tremendously escalated the consumption for biopesticides in the country over the years resulting in a double digit CAGR growth (https://www.researchandmarkets.com/reports/5003583/biopesticide-market-trends-forecast-and#rela3-5214644). In FY'2019 the revenue generated through microbial pesticides consisted of a microorganism (e.g., a bacterium, fungus, virus or protozoan) as the active ingredient which contributed to a majority of the proportion in the overall Biopesticides market. Microbial biopesticides are eco-friendly pests management solutions and have high specificity due to which share of microbial biopesticides has contributed highest share in terms of revenue in FY'2019. Invertebrate pathogenic microorganisms employed as active substances in pest management are recognized as generally safe for the environment and non-target species, in comparison with synthetic chemicals. Botanical/biochemical and PIP was observed to capture the remaining volume share in the FY' 2019.

With improved seed vigor and introduction of systematic disease resistance, the demand for this fungal symbiont has remained high and rendered a majority share in the Indian biofungicides market in FY'2019. During FY'2019, *Bacillus thuringiensis* var. kurstaki contributed to a majority of the bioinsecticides sold in India. Due to its high effectiveness and quicker results, it is preferred over any other bioinsecticide present in the Indian market and thus, contributed the highest share of in the Indian Bioinsecticide market in FY'2019. The application of biopesticides is spread across several crops in agriculture. The share of cereals, pulses and oilseeds has been recorded the highest, and has commanded a major portion of the overall bio-pesticide consumption in India during FY'2019. Flower, spices and tea constituted the smallest market share in terms of revenue generated.

In India, the demand for indigenous biopesticides has dominated the overall biopesticides market during FY'2019. Imported biopesticides which mainly includes Bacillus, semiochemicals and others have accounted for the rest of the market. The consumption of biopesticides was dominated by western region in FY'2019 followed by South, East, North and North Eastern region. The Major companies in the market of biopesticides are EID Parry, T Stanes, Fortune Biotech, Excel Crop Care, International Panaacea Ltd, Biotech International, Kan Biosys, PCI, PJ Margo, Prathibha Biotech and Zytex Biotech. Price, quality and distribution network are some of the critical parameters on the basis of which companies compete in the organized segment.

India Biopesticide Market

In India, the usage of biopesticides is growing at a faster pace than that of the chemical pesticides. According to the Directorate of Plant Protection, Quarantine and Storage, Ministry of Agriculture and Farmer Welfare, in the last 10 years, consumption of bio-pesticides increased by 23%, while that of chemical pesticides grew only by 2%. The total demand by various states/UTs of India was reported as 59,458 MT technical grade of biopesticides which was 8795 MT at 2014-15 and 10,852 MT at 2019-20 with a sharp decline at 2018-19 (9725 MT) (Fig. 1). However, there was a gap between demand and consumption of biopesticides in India in which only 70.2 % of total demand was consumed under various crops, grown in different states of India (Fig. 2,3).

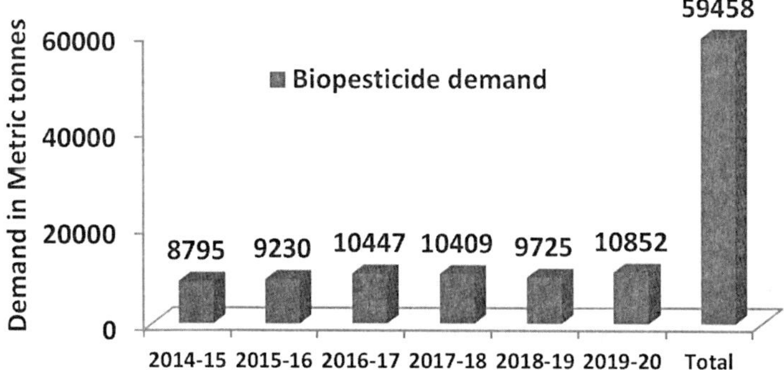

Fig. 1: India biopesticide demand during 2014-2020

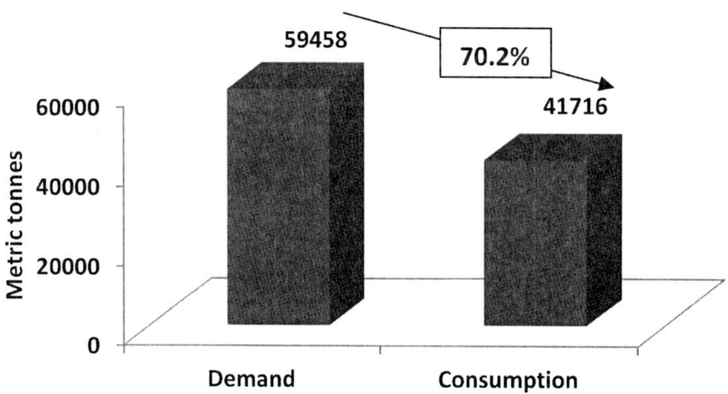

Fig. 2: India biopesticide formulations consumption during 2014-2020

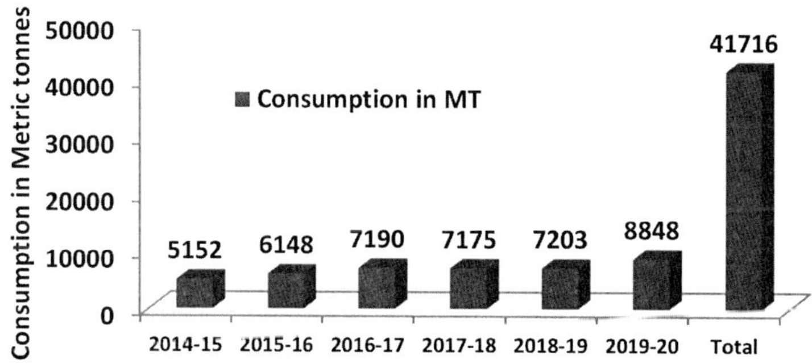

Fig. 3: Total demand and consumption of biopesticide formulations in India during 2014-2020

The total consumption of biopesticides was computed as 22,404 MT technical grade during 2014-2020 which was 9.07% of total chemical pesticides consumed (2,41,969 MT) (Fig. 4).

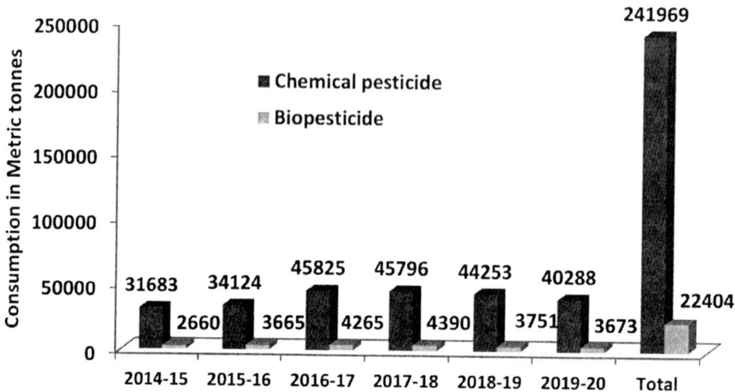

Fig. 4: Comparative analysis on consumption of chemical pesticides and biopesticides in India during 2014-2020

Among the states, Maharashtra was leading to absorb maximum quantity of biopesticides (6630 MT) during 2014-2020, followed by West Bengal (5433 MT), Kerala (4257 MT), Karnataka (3160 MT), Tamil Nadu (2816 MT), Madhya Pradesh (2737 MT), Chhattisgarh (2549 MT) and Haryana (2250 MT) (Fig. 5) (https ://ppqs.gov.in/statistical-database).

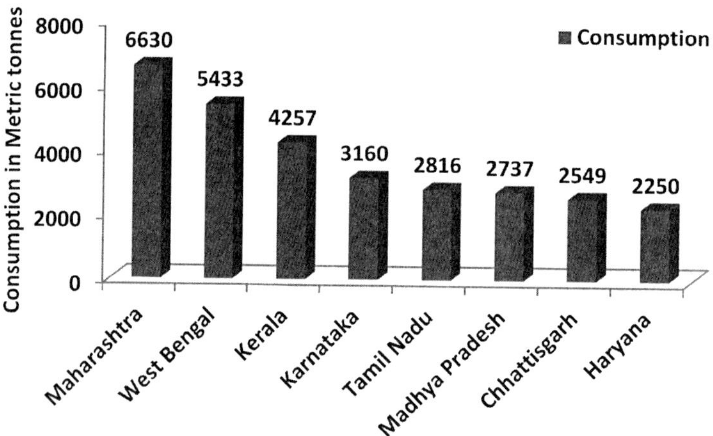

Fig. 5: State-wise consumption of biopesticide formulations during 2014-20 in India

The 9.4% of total cultivated areas (4.86 million ha) in cereals, pulses, oilseeds, fibre, furits, vegetables, plantation and other crops during 2014-2020 was reported to be covered with biopesticides in India. Cereal crops consumed the maximum quantity of biopesticides (4876 MT), followed by oilseeds (4434 MT), vegetables (3980 MT), pulses (2905 MT), fruits (1819 MT) etc., during 2014-2020 (Fig. 6)

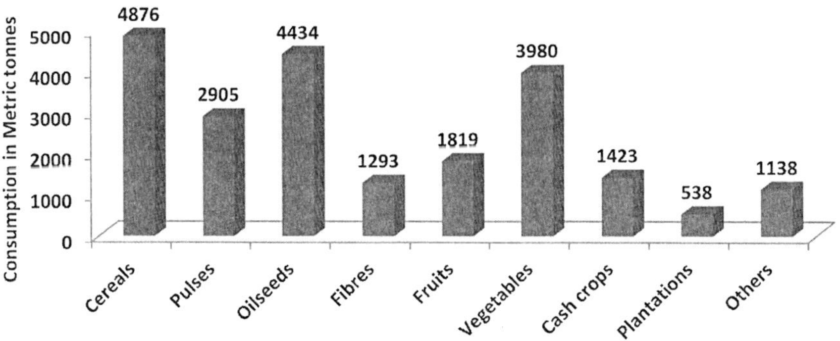

Fig. 6: Crop-wise consumption of biopesticide formulations during 2014-20 in India

Global Biopesticide Market Growth

Global biopesticides market has occupied in a small fraction of the total global crop protection market and it was estimated as $ 1.72 billion in 2014; $ 1.89 billion in 2015; $ 2.09 billion in 2016; $ 2.31 billion in 2017; $ 2.55 billion in 2018; $ 2.81 billion in 2019; and $ 3.09 billion in 2020. The biopesticide market is anticipated to contribute $ 3.42 billion in 2021; $ 3.77 billion in 2022; $ 4.5 billion in 2023; and $ 8.19 billion in 2025 (Fig. 7). At 2020, the compound annual growth rate (CAGR) of global biopesticide market was approximately 3-5% of the total crop protection market (Marrone 2014; Olson, 2015; Kumar et al. 2018; Damalas and Koutroubas 2018) while the market was anticipated to grow by 8.64 % at 2023; 9.7% at 2015-2023; 10.3% at 2014-2022; 15% at 2019-2024; 16% at 2020-2025 (https://www.researchandmarkets.com/reports/5003583/biopesticide-market-trends-forecast-and#rela3-5214644).

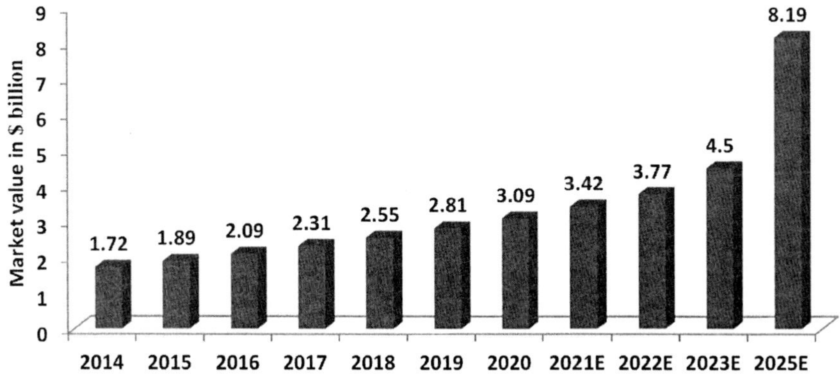

Fig. 7: Global Biopesticide Market Value during 2014-2025

India Biopesticide Market Growth

The India biopesticides market generated revenue of $102 million in 2016 and is anticipated to contribute $778 million by 2025, growing at a CAGR of 25.4% (Fig. 8). The growth rate of biopesticide market in India varied in different periods of report and the CAGR was reported/ anticipated to be 9.3% at 2013-2018; 16.4% at 2013-2019; 7.3% at 2016-2026; 25.4% at 2017-2025; 10.3% at 2018-2024; 13.1% at 2019-2027; 25.1% at 2016-2025 (https:// inkwoodresearch.com/reports/india-biopesticides-market-forecast- 2017-2025) (Fig. 8).

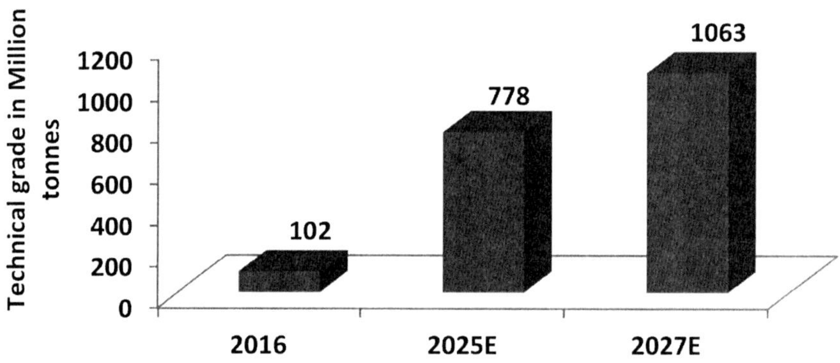

Fig. 8: India Biopesticde market value during 2016-2027

Conclusions

As restrictions have tightened in recent years, the pipeline of new chemistry has significantly decreased. Products are being pulled off the market because they no longer adhere to the severe standards. As a result, a smaller selection of chemical remedies continues to target numerous pests in a small number of staple crops. These effects, which have always been clear in the market for pesticides, are now more clear than ever. Future market growth for biopesticides will be closely correlated with biological control agent research. There are few comprehensive and systematic studies on the preliminary research that a number of scientists from various research institutes have conducted in the area. Therefore, it is crucial to improve businesses and research institutions' cooperation on this issue. The agriculture industry can and should profit from the coexistence of biopesticides and chemical pesticides as it appears that biopesticides cannot yet totally replace chemical pesticides. In this context, it is anticipated that large-scale industrial development will be facilitated by speeding the practical application of research findings.

References

Balog A, Bartel T, Loxdale HD, Wilson K (2017) Differences in the progress of the biopesticide revolution between the EU and major crop-growing regions. Pest Manag Sci 73:2203-2208.

Czaja K, Goralczyk K, Strucifnski P, Hernik A, Korcz W, Minorczyk M, Lyczewska M, Ludwicki JK (2015) Biopesticides-Towards increased consumer safety in the European Union. Pest Manag Sci 71:3-6.

Damalas CA, Koutroubas SD (2018) Current status and recent developments in biopesticide use. Agriculture 8(13). doi:10.3390/agriculture8010013.

Kumar KK, Sridhar J, Murali-Baskaran RK, Senthil-Nathan S, Kaushal P, Dara SK, Arthurs S (2019) Microbial biopesticides for insect pest management in India: current status and future prospects. J Invert Pathol 165:74-81.

Kumar S, Singh A (2015) Biopesticides: Present status and the future prospects. J Fertil Pestic 6e 129.

Marrone PG (2014) The market and potential for biopesticides 245-258pp. In: AD Gross et al. (eds), Biopestiides: State of the Art and Future Opportunities, American Chemical Society, Washingtone DC, USA.

Olson S (2015) An analysis of the biopesticide market now and where is going. Outlooks Pest Manag 26:203-206.

Links

https ://ppqs.gov.in/statistical-database

https://inkwoodresearch.com/reports/india-biopesticides-market-forecast-2017-2025

https://www.researchandmarkets.com/reports/5003583/biopesticide-market-trends-forecast-and#rela3-5214644

https://www.researchandmarkets.com/reports/5003583/biopesticide-market-trends-forecast-and#rela3-5214644

8

Biopesticides Research and Development Activities

Abstract

In the creation of prospective biopesticides, bacteria and fungi are the main focus among the beneficial microbes utilised in biological control. Even if there is more research and development being done on the discovery of microorganisms and subsequent commercialization, the slow kill rate, short shelf life, resistance to microbe etc., are seen as a setback in the biopesticide industry. Several attempts have been made to increase the rate of kill, including developing Bt-based transgenics, transferring the genes encoding the production of toxins from scorpion, spider, etc. into the genome of helpful microbes etc., that increased the rate of kill by several folds. To perfect such genetically modified microbe based biopesticides, additional research is necessary.

Keywords: Beneficial microbes, Biopesticide, Genetic improvement

Introduction

Worldwide, various biopesticides have been created and are in use, including viruses, microorganisms (bacteria, fungi, etc.), microorganism derived products, animal derived goods (pheromones, hormones, insect-specific toxins, etc.), plant derived products, and genetically modified organisms (Nicholson 2007; Erlandson 2008; Mazhabi et al. 2011; Islam and Omar 2012). Microbial biopesticides are the greatest class of broad-spectrum, pest-specific among all currently employed biopesticides (safe to non-target organisms and eco-friendly). Thirty member nations of the Organization for Economic Co-operation and Development (OECD) offer more than 200 microbial biopesticides (Kabaluk and Gazdik 2007). According to Kiewnick (2007), there are 21 microbial biopesticides registered in the European Union (EU), 22 in Canada, and 53 in the United States. However, reports of the items registered for usage in Asia vary (Thakore 2006). Overall, microbial biopesticide registrations are increasing globally, the expansion of various technologies has increased the scope for more products and the change in the trend to develop microbial products is definitely on the rise (Bailey et al. 2010 and Kristiofferesen et al. 2008; Shukla 2019).

Entomopathogenic Bacteria

Bacillus that forms crystalliferous spores (*Bacillus thuringiensis*), obligate pathogens (*Bacillus popilliae*), prospective pathogens (*Serratia marcesens*), and facultative pathogens (*Pseudomonas aeruginosa*) can all be classified as biopesticide-producing bacteria. Due to their efficacy and safety, spore formers have been the most commonly used for commercial purposes. *Bacillus sphaericus* and *B. thuringiensis* are the most often used microorganisms. A unique, secure, and efficient tool for controlling insects is *B. thuringiensis* (Roy et al. 2007). It is largely a pathogen of lepidopterous pests like rice stem borers and the American bollworm in cotton. *Bt* releases poisons when consumed by insect larvae, damaging the pest's midgut and ultimately killing it. The strains of the subspecies kurstaki, galeriae, and dendrolimus are the primary sources for the manufacturing of *Bt* preparations. Other bacterial species have limited effect on pest management, although there are commercial products based on *Agrobacterium radiobacter, B. popilliae, B. subtilis, Pseudomonas cepacia, Pseudomonas chlororaphis, Pseudomonas flourescens, Pseudomonas solanacearum,* and *Pseudomonas syringae.*

Insect Viruses

More than 700 viruses that infect insects have been discovered, the majority of which originate from Lepidoptera (560), followed by Hymenoptera (100), Coleoptera, Diptera, and Orthoptera (40) (Khachatourians 2009). A dozen or more of these viruses have been made available for commercial usage as biopesticides. The RNA-containing reoviruses, cytoplasmic polyhedrosis viruses, nodaviruses, picrona-like viruses, and tetraviruses, as well as the DNA-containing baculoviruses (BVs), nucleopolyhedrosis viruses (NPVs), granuloviruses (GVs), acoviruses, iridoviruses, parvoviruses, polydnaviruses, and poxviruses are used in insect management. However, NPVs and GVs have been the primary categories utilised in pest management. These viruses are efficient against insects that consume plants and are used extensively around the world to control pests in vegetables and field crops. Their use has significantly reduced the populations of gypsy moths, pine sawflies, Douglas fir tussock moths, and pine caterpillars in forest settings. Potato tuber moth is controlled by *Phthorimaea operculella* GVs in stored tubers, and codling moth is managed by *Cydia pomonella* GVs on fruit trees (Arthurs et al. 2008). Insects including cabbage moths, corn earworms, cotton leaf worms and bollworms, beet armyworms, celery loopers, and tobacco budworms can also be controlled with virus-based solutions. Target-specific viruses called baculoviruses can infect and kill a variety of significant plant pests. When they are used against lepidopterous pests of cotton, rice, and vegetables, they are especially effective. Their use has been constrained to small areas because of the challenges associated with their large-scale manufacture. They are not available commercially in India, but are being developed on a modest scale by various IPM institutions and state agricultural departments. Natural baculoviruses have been successfully used to preserve crops and forests, but from an agro-industrial standpoint, they are ineffective insecticides and have a number of potential drawbacks (Possee et al. 1997; Inceoglu et al. 2006). Compared to chemical insecticides, they have a slower rate of mortality (from five days to more than two weeks) and have a narrower host specificity, limited field stability, susceptibility to UV exposure, short shelf life, and higher production costs.

Entomopathogenic Fungi

Trichoderma harzianum, Trichoderma viridae, Streptomyces griseoviridis, Verticillium chlamydosporium, Beauveria bassiana, Metarhizium anisopilae, Nomuraea rileyi, Paecilomyces farinosus, and *Verticillium lecanii* are some of the most often employed species and many of them have received global commercialization. An efficient fungicide against root rot that is transmitted through the soil is *Trichoderma*. It is especially important for dry land crops like chickpeas, groundnuts, black gram, and green gram that are prone to various diseases. *Trichoderma* based biopesticide is simple to make and only needs a fundamental understanding of microbiology. For the management of soil- and seed-borne diseases, this bio-fungicide is advised for use as a seed treatment, soil application, soil drenching, root dip technique, etc. Important crop diseases which are well managed with *Trichoderma* based biopesticides are *Armillaria, Botrytis, Chondrostereum, Colletotrichum, Dematophora, Diaporthe, Endothia, Fulvia, Fusarium, Fusicladium, Helminthosporium, Macrophomina, Monilia, Nectria, Phoma, Phytophthora, Plasmopara, Pseudoperonospora, Pythium, Rhizoctonia, Rhizopus, Sclerotinia, Sclerotium, Venturia, Verticillium,* and wood rot fungi. Many *Trichoderma* strains, mainly *T. harzianum*, *T. viride* and *T. virens* (formerly *Gliocladium virens*) play vital role in plant diseases management (Singh 2014). Additionally, recent studies suggest that Trichoderma strains may be used to handle abiotic stresses as salt and drought (Shukla et al. 2012; Rawat et al. 2011). In addition, green muscardine fungus (*Metarizhium anisopliae*), white halo fungus (*Beauveria bassiana*) and *verticillium lecanii* based biopesticides are also popular in management of beetle pests, Lepidoptera pests, sucking pests etc.

Entomopathogenic Nematodes

The entomopathogenic nematodes (EPN), which control weevils, gnats, white grubs, and numerous species of the Sesiidae family, are another group of biopesticide (Klein 1990; Shapiro-Ilan et al. 2002; Grewal 1990). Insects feeding in enigmatic settings such as soil-borne pests and stem borers are kept under control by this interesting EPN. Nematodes from the genera Steinernema and Heterorhabditis, which attack hosts as infective juveniles (IJs), are frequently used in pest management (Kaya and Gaugler 1993; Koppenhofer and Kaya 2002).

Protozoans

The use of protozoan pathogens as biopesticide agents has not been particularly effective, despite the fact that they naturally infect a wide variety of pests and cause chronic and crippling effects that lower the target pest populations. Taxonomically speaking, protozoa are split into various phyla, some of which have entomogenous species. In-depth research has been done on microsporan protozoans as potential inclusions in integrated pest management plans. For many insect species, microsporidia are the disease-causing intracellular parasites that are ubiquitous and necessary. Because they target lepidopteran and orthopteran insects and tend to kill hoppers more frequently than any other insect, two genera, Nosema and Vairimorpha, offer some potential (Lewis 2002). According to research on the microsporidium *Nosema pyrausta*, which infects the European corn borer *Ostrinia nubilalis*, a spore is consumed by a larva of the European corn borer, which then germinates in the midgut, extrudes a polar filament, and injects sporaplasm into a midgut cell. The sporaplasm multiplies and creates additional spores, which can spread infection to other tissues. Infected midgut cells shed their spores into the gut lumen, where they are eliminated to the maize plant with the animal's waste. The infection cycle is repeated in the midgut cells of the new host as a result of these spores, which are still viable, being ingested during larval feeding. If a female larva is affected, Nosema is vertically transmitted to the filial generation. The developing oocytes and ovarian tissue get infected with *N. pyrausta* as the infected larva grows into an adult. When the larvae hatch, they are infected with *N. pyrausta* since the embryo is already contaminated within the yolk. *N. pyrausta* is maintained in naturally occurring populations of the European corn borer by both horizontal and vertical transmissions.

Resistance to Microbes

The development of resistance has been observed most frequently in *B. thuringiensis* among the numerous families of microbial pathogens. At least 16 insect species have been discovered recently that are resistant to *B. thuringiensis*. Noctuid species like *Spodoptera frugiperda*, *Busseola fusca*, and *H. zea* have been found to have developed resistance to 8-endotoxins in the field (Tabashnik et al. 2009). The majority of reports of the development of resistance in *Plutella xylostella* field populations come from the nations that employ *Bacillus thuringiensis* widely, including China, Japan, the Phillipines, Malaysia, India, and North America. We now have *B. thuringiensis Bt* cotton and *B. thuringiensis* maize available in 13 and nine countries, respectively, grown on 42.1 million ha of land (Shelton et al. 2008). Genetic engineering was thought to be a useful tool to avoid this resistance problem where microbial genes from *B. thuringiensis* were transferred to plants to produce transgenics. In terms of microbial pest control, the introduction of such transgenics was hailed as a miracle cure; nevertheless, field resistance in *H. zea* as a result of an increase in the frequency of resistance alleles is concerning (Tabashnik et al. 2008). The field-evolved insect resistance to *B. thuringiensis* crops and various aspects related to resistance monitoring methods have been comprehensively reviewed recently (Tabashnik et al. 2009); obviously more prominent in lepidopterans (Downes et al. 2010; Huang et al. 2011). Factors associated with field resistance are the failure to use high dose *B. thuringiensis* cultivars and lack of a sufficient refuge. While implementation of the high-dose/refuge insect resistance management strategy has been successful in delaying field resistance to *Bt* crops (Huang et al. 2011), Gene pyramiding is another approach used to try and address the emerging resistance problem (Zhao et al. 2003; Manyangariwa et al. 2006). Pyramiding is the stacking of various genes to cause the transgenic plant to express numerous toxins. However, gene pyramiding must be sustained and shouldn't result in numerous resistances or cross-resistances. Multiple resistance cannot be completely ignored because doing so would render these techniques useless in the end. In order to maintain the efficacy of pyramided *B. thuringiensis* crops, it is critical to account for the potential implications of such cross-resistance in resistance management plans. Pink bollworm has asymmetrical cross-resistance between *B. thuringiensis* toxins Cry1Ac and Cry2Ab (Tabashnik et al. 2009).

Gene pyramiding may not be a sustainable tactic per se, according to recent research, thus management plans must also include other tactics including refugia, the employment of predators and parasitoids, and crop rotation schemes (Zhao et al. 2003; Tabashnik et al. 2009). Soon, RNA interference-based transgenic plants that control insects will be a reality (Baum et al. 2007; Mao et al. 2007), expanding the potential applications of transgenics and reducing the negative effects of resistance. Recent research has demonstrated that toxin-binding proteins like *cadherin* increase the toxicity of *B. thuringiensis* (Soberon et al. 2007). In contrast to the usual *B. thuringiensis* toxins, these binding proteins help toxin oligomerization and hence change the toxin, which can avoid resistance. The experiments show that *cadherin* gene silencing using RNA interference in *M. sexta* reduces the toxicity of *B. thuringiensis* toxin Cry1Ab. *M. sexta* and *Pectinophora gossypiella* that were resistant to *B. thuringiensis* were killed by the toxins that possessed *cadherin* deletion mutations (Soberon et al. 2007).

Recently, resistance in a baculovirus in the field has been found in Europe where *Cydia pomnella* GV is one of the main components of the codling moth control. *C. pomonella* GV in apple orchards has led to a high degree of resistance in some populations (Sauphanor et al. 2006; Frisch et al. 2007). This is the first documented instance of field resistance to a commercially applied baculovirus (Eberle and Jehle 2006). Apparently, this is either the result of the overuse of the product or the predominant control strategy applied. However, there do not seem to be any reported examples of field development of resistance to entomopathogenic fungi or nematodes (Shelton et al. 2007). However, there is evidence to demonstrate the existence of natural resistance mechanisms in insects against fungi (Wilson et al. 2001) and nematodes (Kunkel et al. 2004), suggesting that resistance to these pathogens cannot be summarily ignored.

Genetic Improvement of Insect Pathogens

Entomopathogenic Bacteria

The goal of genetically modifying microbial pathogens is to increase their potential to cause disease by enhancing toxin production, reproduction, and transmission rates. One strain of *B. thuringiensis*, for instance, exhibits insecticidal action against both coleopteran and lepidopteran insects as a result of genetic modification (Lereclus et al. 1992). Genetic modification can potentially increase *B. thuringiensis* activity on crop foliage or in soil treatments. For instance, the Cry34 and Cry35 families of crystal proteins from *B. thuringiensis* operate as binary toxins with action against the western maize rootworm, *Diabrotica virgifera virgifera*. Pairings Cry34A/Cry35A are busier than pairs Cry34B/Cry35B. The binary Cry34/Cry35 *B. thuringiensis* crystal proteins are closely linked to one another, are found throughout the environment, and have sequence similarities that are consistent with their ability to affect their target organisms' membranes. Plant pests and rootworms can be effectively controlled by modified Cry35 proteins, which have had their segments, domains, and motifs swapped with those of other proteins to increase their insecticidal activity (Schnepf et al. 2007). Similar to this, the *B. thuringiensis* Cry8Bb1 toxin polypeptide was developed to feature a proteolytic protection site that renders it insensitive to a plant protease, aiding in the toxin's protection from any proteolytic inactivation. Modified Cry8Bb1 has been used for controlling corn rootworms, wireworms, boll weevils, Colorado potato beetles and the alfalfa weevils (Abad et al. 2008).

A new study demonstrates the presence of the *Bacillus enhancin-like* (*bel*) gene in the genomes of the *B. cereus* group, which has the potential to boost the insecticidal action of biopesticides based on *B. thuringiensis* and transgenic plants derived from *B. thuringiensis* genes (Fang et al. 2009). Bel genes produce peptides that resemble viral enhancin protein by 20-30%. As they destroy the peritrophic matrix of insect midguts, these proteins are known to strengthen viral infections. The mortality rate was 2.2 times higher when *Bel* and *Cry1Ac* were combined (Fang et al. 2009).

Insect Baculoviruses

Since the slow mortality rate of wild-type baculoviruses makes them impractical to utilise, numerous methods have been devised to increase the baculovirus's ability to kill by introducing genes encoding insect hormones, enzymes, or particular toxins (Kamita et al. 2005; Li and Bonning 2007; Gramkow et al. 2010). Maeda was the first to create a genetically altered baculovirus that expressed a gene encoding a hormone successfully in 1989 (Maeda 1989). This baculovirus produced the gene for a diuretic hormone, which led to water loss in Bombyx mori larvae and interfered with the insect's normal physiology. This modified BmNPV had a 20% quicker kill rate than the parent BmNPV's wild-type counterpart. This work developed a novel idea and laid the groundwork for later usage of baculoviruses to eradicate insects. Other enzymes and hormones were tested to alter baculoviruses in the years that followed.

Recombinant baculoviruses have demonstrated promise as more effective insect pest controllers. However, it is important to consider how utilising such viruses may affect the environment. Baculoviruses are not contagious to non-target creatures, including beneficial insect species, predators, and parasitoids of lepidopteran larvae, according to the findings of several research (Boughton et al. 2003; Sun et al. 2009). According to Hartig et al. (1991), recombinant AcMNPV baculovirus expressing AaiT was not infectious to adherent mammalian cells, and recombinant HaSNPV expressing AaiT was not pathogenic to fish, birds, or other vertebrates in any way (Sun et al. 2002). A recombinant baculovirus does not possess any selected ecological advantages over the wild-type baculovirus, according to numerous investigations conducted both in the field and in greenhouse environments (Cory et al. 1994; Black et al. 1997; Lee et al. 2001). Additionally, there is a negative selection towards recombinant baculoviruses, which causes the wild-type to swiftly displace them (Georgievska et al. 2010; Zwart et al. 2010). The likelihood that the cloned gene will transfer from the recombinant baculovirus to another creature has also been conjectured. Although this is theoretically possible, it has never been demonstrated because of variables that prevent or restrict the occurrence of this genetic recombination (Inceoglu et al. 2001). Combination viruses have the potential to be more effective insect pests.

Recombinant baculoviruses were effectively used to express juvenile hormone esterase (Hammock et al. 1990), eclosion hormone (Eldridge et al. 1991), and prothoracicotropic hormone (O'Reilly et al. 1995). Only those producing juvenile hormone esterase, however, significantly outperformed parent wild-type baculoviruses in terms of insecticidal efficacy (El-Sheikh et al. 2011a). Juvenile hormone esterase controls the hormone, therefore when it is over expressed, the hormone's concentration falls. This causes the insect to stop feeding and pupate (El-Sheikh et al. 2011b). The effective utilisation of recombinant baculoviruses expressing this enzyme is severely hampered by the juvenile hormone esterase's brief half-life in the hemolymph. However, numerous attempts have been made to increase in vivo stability in order to make it more effective (Hinton and Hammock 2003; Inceoglu et al. 2006; Kamita and Hammock 2010).

Baculoviruses that have undergone genetic modification to express toxins have been widely used in the past. The first successful insertion of toxin genes into baculoviruses was reported in the late 1980s (Carbonell et al. 1988; Tomalski et al. 1991; Ooi et al. 1989). Since then, most studies have concentrated on comprehending arthropod-specific venoms produced by mites, spiders, or scorpions (Inceoglu et al. 2006). The *Androctonus australis* insect-specific toxin (AaiT) was the first and most effective insecticide expressed in baculoviruses (MacCutchen et al. 1991; Maeda et al. 1991; Stewart et al. 1991). When compared to the parent wild-type baculovirus, using a recombinant *Bombix mori* baculovirus (BmNPV) expressing AaiT accelerated the death of silkworm larvae by up to 40%. (Maeda et al. 1991). Another study utilising a different baculovirus expressing AaiT revealed that *Manduda sexta* larvae were paralysed many hours before death, increasing pesticidal efficacy (MacCutchen et al. 1991). The efficacy of baculoviruses that express AaiT was further validated in field experiments. (Cory et al. 1994; Sun et al. 2002; Sun et al. 2004).

Although AaiT has been the focus of numerous studies and is thought to be the best model peptide neurotoxin for enhancing the insecticidal activity of baculoviruses (Inceoglu et al. 2006; Sun et al., 2009), other scorpion toxins such as those from *Leiurus quinquestriatus quinquestriatus, Leiurus quinquestriatus hebraeus*, and *Buthus martcn* (Kopeyan et al., 1990; Zlotkin et al. 1993; Moskowitz et al. 1998; Froy et al. 2000; Tang et al. 2011), spiders *Agelenopsis aperta*, *Dighetia canities*, *Tegenaria agrestis* and *Araneus ventricosus* (Prikhodko et al. 1996; Hughes et al. 1997; Jung et al. 2012), or straw itch mite, *Pyemotes tritici* is another source of powerful toxins that, when expressed in baculovirus, are active against insect pests and may one day be employed as biopesticides (Tomalski and Miller 1991; Lu et al. 1996; Burden et al. 2000). Another method for quickening the death of the baculoviruses is to delete an endogenous gene, such as the gene encoding the baculovirus-encoded enzyme ecdysteroid UDP-glucosyltransferase (O'Reilly and Miller 1991). Because ecdysteroids are hormones that regulate larval-pupal moulting and eating behaviour, infection with an egt deletion mutant baculovirus results in a reduction in food consumption and an early mortality (Eldridge et al. 1992; Wilson et al. 2000; Cai et al. 2010; Georgievska et al. 2010).

The insect sodium channel is the molecular target of the majority of these neurotoxins (Cestele and Catterall 2000; Casida and Durkin 2013), which is also the main target of insecticides of the pyrethroid class. However, since their individual binding sites on the channel do not overlap, there is a chance of creating a synergistic effect that would permit the employment of both pyrethroids and baculoviruses that express toxins at the same time (McCutchen et al. 1997). The newest strategy involves the expression of the crystal protein gene from *Bacillus thuringiensis* in *Autographa californica* mNPV. In comparison to the parent wild-type AcMNPV, this recombinant baculovirus has demonstrated a high insecticidal activity against *Spodoptera exigua* and *Plutella xylostella* (Jung *et al.* 2012; Shim *et al.* 2013).

Entomopathogenic Fungi

Metarhizium anisopliae and *B. bassiana*, two commonly employed entomopathogenic fungi, have undergone substantial study for the clarification of pathogenic mechanisms and alteration of the genes of the pathogens to increase biocontrol efficacy (St. Leger et al. 2010). In the genome of *M. anisopliae*, extra copies of the gene encoding the controlled cuticle-degrading protease Pr1 were introduced and over expressed. Compared to the parent wild-type strain, the offspring strain decreased tobacco hornworm (*M. sexta*) survival time by 25%. (St. Leger et al. 1996). The scorpion toxin (AaIT) expressed in the *M. anisopliae* strain ARSEF 549 illustrates the astonishing extent to which pathogenicity can be boosted. At 22-fold lower spore dosages than the wild type, the modified fungus produced the same mortality rates in *M. sexta*, and survival times at some concentrations were 40% lower (Wang et al. 2007). Similar outcomes have been seen with mosquitoes, where the LC_{50} was reduced by 9 times, and coffee berry borer beetles, where the LC_{50} was reduced by 16 times (Pava-Ripoll et al. 2008).

Entomopathogenic Nematode

Artificial selection has proven successful in boosting infectivity and nematicide resistance in entomopathogenic nematodes (Griffin 1993). With relation to host penetration and reproductive potential, the strain selection has demonstrated an improvement in fitness. The possibility of examining whether a selection strategy might enhance the control of root pests has been made possible by the recent revelation that maize roots harmed by the western corn rootworm release a crucial attractant for insect-killing nematodes (Hiltpold et al. 2010). After 10 to 25 selection cycles, a diverse population of *Steinernema feltiae* was produced for desiccation tolerance and host-seeking capacity. Artificial selection for one characteristic, however, may come at the expense of other crucial traits like contagiousness, establishment, and/or field persistence. In the near future, it may be possible to produce GM nematodes with higher storage stability, more resilience to environmental challenges, and greater biological control potential using data from the sequenced genomes of EPN (Sandhu et al. 2006; Ciche et al. 2007; Bai et al. 2009).

Conclusions

Despite having a slower death rate than chemical pesticides, wild-type beneficial organisms have demonstrated to be an effective long-term solution in particular situations, such as forest ecosystems. However, the parent wild-type microbe does not kill insects as quickly as recombinant microbes do, which is a severe drawback. These recombinant microorganisms are generally made of toxin genes from scorpions or spiders. Hopefully, recombinant microorganisms will get more market share worldwide. Numerous papers state that there is no evidence showing that genetically modified organisms pose a greater harm to animals and the environment than do organisms of the natural variety.

References

Abad AR, Flannagan RD, McCutchen BF, Yu CG (2008) *Bacillus thuringiensis* cry gene and protein. US Patent 20087329736.

Arthurs, SP, Lacey LA, Rosa FDL (2008) Evaluation of granulovirus (PoGV) and *Bacillus thuringiensis* subsp. Kurstaki for control of the potato tuberworm (Lepidoptera: Gelechiidae) in stored tubers. J Econ Entomol 101:1540-1546.

Bai X, Adams BJ, Ciche TA, Clifton S, Gaugler R, Hogenhout SA et al (2009) Transcriptomic analysis of the entomopathogenic nematode Heterorhabditis bacteriophora TTO1. BMC Genom 10:205-218.

Bailey, KL, Boyetchko SM, Langle T (2010) Social and economic drivers shaping the future of biological control: a Canadian perspective on the factors affecting the development and use of microbial biopesticides. Biol Control 52:221-229.

Baum JA, Bogaert T, Clinton W, Heck GR, Feldmann P, Ilagan 0, et al (2007) Control of coleopteran insect pests through RNA interference. Nature Biotechnol 25:1322-1326.

Black BC, Brennan, LA, Dierks PM, Gard IE (1997) Commercialization of baculoviral insecticides. In: Miller, L. K. (ed) The Baculoviruses. Plenum, New York, USA, 341-387pp.

Boughton AJ, Obrycki JJ, Bonning BC (2003) Effects of a protease-expressing recombinant baculovirus on nontarget insect predators of *Heliothis virescens*. Biol Control 28:101-110.

Burden J, Hails RS, Windass JD, Suner MM, Cory J (2000) Infectivity, speed of kill, and productivity of a baculovirus expressing the itch mite toxin txp-1 in second and fourth instar larvae of *Trichoplusia ni*. J Invertebr Pathol 75:226-236.

Caballero, P., R. Murillo, D. Muñoz and T. Williams (2009) The nucleopolyhedrovirus of *Spodoptera exigua* (Lepidoptera: Noctuidae) as a biopesticide: analysis of recent advances in Spain. Rev Colom Entomol 35:105-115.

Cai Y, Cheng Z, Li C, Wang F, Li G, Pang Y (2010) Biological activity of recombinant *Spodoptera exigua* multicapsid nucleopolyhedrovirus against *Spodoptera exigua larvae.* Biol Control 55:178-185

Casida JE Durkin KA (2013) Neuroactive insecticides: targets, selectivity, resistance, and secondary effects. Annu Rev Entomol 58:99-117.

Cestele S, Catterall WA (2000) Molecular mechanisms of neurotoxin action on voltage gated sodium channels. Biochimie 82:883-892.

Ciche TA, Sternberg PW (2007) Postembryonic RNAi in Heterorhabditis bacteriophora: A nematode insect parasite and host for insect pathogenic symbionts. BMC Dev Biol 7:101-111.

Cory JS, Hirst ML, Williams T, Hails RS, Goulson D, Green BM, Carty TM, Downes RDS, Mahon RJ, Rossiter L, Kauter G, Leven T, Fitt G, (2010) Adaptive management of pest resistance by *Helicoverpa* species (Noctuidae) in Australia to the Cry2Ab Bt toxin in Bollgard II cotton. Evol Appl 3:574-584.

Eberle KE, Jehle JA (2006) Field resistance of codling moth against *Cydia pomonella* granulovirus (CpGV) is autosomal and incompletely dominant inherited. J Invertebr Pathol 93:201-206.

Eldridge, R., F. M. Horodyski, D. B. Morton, D. O'Reilly, J. W. Truman, L. M. Riddiford and L. K. Miller (1991) Expression of an eclosion hormone gene in insect cells using baculovirus vectors. Insect Biochem. 21: 341-51.

El-Sheikh E.S.A., S. G. Kamita, K. Vu and B. D. Hammock (2011a) Improved insecticidal efficacy of a recombinant baculovirus expressing mutated JH esterase from *Manduca sexta*. Biol Control 58:354-361.

El'Sheikh ESA, Mamtha MD, Ragheb DA, Ashour BA (2011b) Potential of juvenile hormone esterase as a bio-insecticide: an overview. Egyp J Biol Pest Cont 21(1):103-110

Erlandson ME (2008) Insect pest control by viruses. In: BWJ Mahy, MHV Van REgenmortel (eds), Encyclopedia of virology, 3rd ed Elsevier, Oxford 125-133pp.

Fang S, Wang L, Guo W, Zhang X, Peng D, Luo C, (2009) *Bacillus thuringiensis* Bel Protein enhances the toxicity of Cry1Ac protein to *Helicoverpa armigera* larvae by degrading insect intestinal mucin. Appl Environ Microbiol 75:5237-5243.

Fowler (1991) Insecticidal effects of an insectspecific neurotoxin expressed by a recombinant baculovirus. Virol 184:77-80.

Frisch E, Undorf-Spahn K, Kienzle J, Zebitz CPW, Huber J (2010) Codling moth granulosivirus: First indication of variations in the susceptibility of local codling moth populations. IOBC/from the essential initiation site of a baculovirus polyhedrin gene. J Mol Biol 210:721-736.

Froy O, Zilberberg N, Chejanovsky N, Anglister J, Loret E, Shaanan B, Gordon D, Gurevitz M (2000) Scorpion neurotoxins: structure/function relationship and application in agriculture. Pest Manag Sci 56:472-474.

Georgievska L, Joosten N, Hoover K, Cory JS, Vlak JM, van der Werf W (2010) Effects of single and mixed infections with wild type and genetically modified *Helicoverpa armigera* nucleopolyhedrovirus on movement behaviour of cotton bollworm larvae. Entomol Exp Appl 135:56-67.

Gramkow AW, Perecmanis S, Sousa RLB, Noronha EF, Felix CR, Nagata T, Ribeiro BM (2010) Insecticidal activity of two proteases against *Spodoptera frugiperda* larvae infected with recombinant baculoviruses. Virol J 29(7):143.

Grewal PS, Ehlers RU, Shapiro-Ilan DI (2005) Nematodes as Biological Control Agents. CAB International, Wallingford, UK.

Griffin CT (1993) Temperature responses of entomopathogenic nematodes: Implications for the success of biological control programmes. In: R. Bedding, R Akhurst, HK Kaya (eds), Nematodes and the Biological Control of Insects. CSIRO, East Melbourne 115-126pp.

Hammock BD, Bonning BC, Possee RD, Hanzlik T, Maeda S (1990) Expression and effects of the juvenile hormone esterase in a baculovirus vector. Nature 344:458-461.

Hartig PC, Cardon MC, Kawanishi CY (1991) Insect virus: Assays for viral replication and persistence in mammalian cells. J Virol Methods 31:335-344.

Hiltpold I, Baroni M, Toepfer S, Kuhlmann U, Turlings TC (2010) Selection of entomopathogenic nematodes for enhanced responsiveness to a volatile root signal helps to control a major root pest. J Exp Biol 213:2417-2423.

Hinton AC, Hammock BD (2003) In vitro expression and biochemical characterization of juvenile hormone esterase from *Manduca sexta*. Insect Biochem Mol Biol 33:317-329.

Huang F, Andow DA, Buschman LL (2011) Success of the high-dose/refuge resistance management strategy after 15 years of Bt crop use in North America. Entomol Exp Appl 140:1-16.

Hughes PR., H. A. Wood, J. P. Breen, S. F. Simpson, A. J. Duggan and J. A. Dybas (1997) Enhanced bioactivity of recombinant baculoviruses expressing insect-specific spider toxins in Lepidoptera crop pests. J Invertebr pathol 69:112-118.

Inceoglu AB, Kamita SG, Hammock BD (2006) Genetically Modified Baculoviruses: A Historical Overview and Future Outlook. Adv Virus Res.68: 23-360.

Inceoglu AB, Kamita SG, Hinton AC, Huang Q, Severson TF, Kang KD, Hammock BD (2001) Recombinant baculoviruses for insect control. Pest Manag Sci 57:981-987.

Islam MT, Omar DB (2012) Combined effect of *Beuveria bassiana* with neem on virulence of insect in case of two application approaches. J Anim Plant Sci 22(1):77-82.

Jung MP, Choi JF, Tao XY, Jin BR, Je YH, Park HH (2012) Insecticidal activity of recombinant baculovirus expressing both spider toxin isolated from *Araneus ventricosus* and *Bacillus thuringiensis* crystal protein fused to a viral polyhedrin. Entomol Res 42(6):339-346.

Kabaluk T, Gazdik K, (2007) Directory of microbial biopesticides for agricultural crops in OECD countries. http://www4.agr.gc.ca/ resources/prod /doc/pmc/ pdf/ micro.e.pdf.

Kamita SG, Hammock BD (2010) Juvenile hormone esterase: biochemistry and structure. J Pest Sci 35:265-274.

Kaya HK, Gaugler R (1993) Entomopathogenic nematodes. Annu Rev Entomol 38:181-206

Khachatourians GG (2009) Insecticides, microbials. Appl Microbiol Agro/ Food 95-109pp.

Kiewnick S (2007) Practicalities of developing and registering microbial biological control agents. CAB Reviews: Perspectives in Agriculture Veterinary Science Nutrition and Natural Resources, No.013, CABI Publishing (http://www.cababstracts plus.org/ cabreviews)

Klein MG (1990) Efficacy against soil-inhabiting insect pests. In: Gaugler R, Kaya HK, editors. Entomopathogenic Nematodes in Biological Control. CRC Press, Boca Raton, FL 195-214pp.

Kopeyan C, Mansuelle P, Sampieri F, Brando T, Bahraoui EM, Rochat H, Granier C (1990) Primary structure of scorpion anti-insect toxins isolated from the venom of *Leiurus quinquestriatus*. FEBS Lett 261:423-426.

Koppenhofer AM, Kaya HK (2002) Entomopathogenic nematodes and insect pest management. In: O Koul, GS Dhaliwal (eds), Microbial Biopesticides. Taylor & Francis, London 277-305pp.

Kristiofferesen P, Rask AM, Grundy AC, Franken I, Kempenaar C, Raison J (2008). A review of pesticide policies and regulations for urban amenity areas in seven European countries. Weed Research; 48: 201-14.

Kunkel BA, Grewal PS, Quigley MF (2004) A mechanism of acquired resistance against an entomopathogenic nematode by *Agrotis ipsilon* feeding on perennial ryegrass harboring a fungal endophyte. Biol Control 29:100-8

Lee Y, Fuxa JR, Inceoglu AB, Alaniz SA, Richter AR, Reilly LM, Hammock BD (2001) Competition between wild-type and recombinant nucleopolyhedroviruses in a greenhouse microcosm. Biol Control 20: 84-93.

Lereclus D, Val lade M, Chaufaux J, Arnates 0, Rambaud S (1992) Expansion of insecticidal host range of *Bacillus thuringiensis* by *in vivo* genetic recombination. Bio/technology 10:418-421.

Lewis LC (2002) Protozoan Control of D Pimental (ed), editor. Encyclopedia of Pest Management. Taylor & Francis, UK 673-676pp.

Li H, Bonning BC (2007) Evaluation of insecticidal efficacy of wild-type and recombinant baculoviruses. Mothod Mol Biol 388:379-404.

Lu A, Seshagiri S, Millar LK (1996) Signal sequence and promoter effects on the efficacy of toxin-expressing baculoviruses as biopesticides. Biol Control 7:320-332.

Maeda S, Volrath SL, Hanzlik TN, Harper SA, Majima K, Maddox DW, Hammock BD, Flower E (1991) Insecticidal effects of an insect specific neurotoxin expressed by recombinant baculovirus. Virol 184:77-80.

Manyangariwa W, Turnbill M, McCutcheon GS, Smith JP (2006) Gene pyramiding as a Bt resistance management strategy: how sustainable is this strategy. African J Biotechnol 5:781-785.

Mao YB, Cai WJ, Wang JW, Hong GJ, Tao XY, Wang LJ, et al (2007) Silencing a cotton bollworm P450 monoxygenase gene by plant-mediated RNAi impairs larval tolerance of gossypol. Nature Biotechnol 25:1307-13.

Mazhabi M, Nemati H, Rouhani H, Tehranifar A, Moghadam EM, Kaveh H, Rezaee A (2011) The effect of *Trichoderma* on polianthes qualitative and quantitative properties. J Anim. Plant Sci 21(3):617-621.

McCutchen BF, Hoover K, Preisler HK, Betana MD, Herrmann R, Robertson JL, Hammock BD (1997) Interaction of recombinant and wild-type baculoviruses with classical insecticides and pyrethroid-resistant tobacco budworm (Lepidoptera: Noctuidae). J Econ Entomol 90: 1170-1180.

Merryweather PJ, Weyer U, Harris MP, Hirst M, Booth M, Booth T, Possee RD (1990) Construction of genetically engineered baculovirus insecticides containing the *Bacillus thuringiensis* subp. kurstaki HD-73 delta endotoxin. J Gen Virol 71:1535-1544.

Moskowitz H, Herrmann R, Jones AD, Hammock BD (1998) A depressant insect-selective toxin analog from the venom of the scorpion *Leiurus quinquestriaus hebraeus*. Eur J Biochem 254:44-49.

Nicholson GM (2007) Fighting the global pest problem: Preface to the special Toxicon issue on insecticidal toxins and their potential for insect pest control. Toxicon 49:413-422.

O'Reilly DR, Miller LK (1991) Improvement of a baculovirus pesticide by deletion of the egt gene. Biotechnol 9:1086-1089.

Ooi BG, Rankin C, Miller LK (1989) Downstream sequences augment transcription from the essential initiation site of a baculovirus polyhedron gene. J Mol Biol 210:721-736.

Pava-Ripoll M, Posada FJ, Momen B, Wang CS, St. Leger RS (2008) Increased pathogenicity against coffee berry borer, *Hypothenemus hampei* (Coleoptera: Curculionidae) by *Metarhizium anisopliae* expressing the scorpion toxin (AaIT) gene. J Invert Pathol 99:220-226.

Possee PJ, Cayley and Bishop DHL (1994) Field trial of a genetically improved baculovirus insecticide. Nature 370:138-140.

Possee RD, Barnett AL, Hawtin RE King LA (1997) Engineered baculoviruses for pest control. Pestic Sci 51:462-470.

Prikhodko GG, Robson M, Warmke J, Cohen CJ, Smith MM, Wang P, Warren V, KaczorowskiG, van der Ploeg LHT, Miller LK (1996) Properties of three baculoviruses expressing genes that encode insect-selective toxins: u-Aga-IV, As II, and Sh I. Biol. Control 7:236-244.

Rawat L, Singh Y, Shukla N, Kumar J (2011) Alleviation of the adverse effects of salinity stress in wheat (*Triticum aestivum* L.) by seed biopriming with salinity tolerant isolates of *Trichoderma harzianum*. Plant soil 347(1-2):387.

Regenmortel (2008) Encyclopedia of virology, 3rd ed. Elsevier, Oxford 125-133pp. ISBN-10: 0123739357

Roh JY, Choi JY, Li MS, Jin BR, Je YH (2007) *Bacillus thuringiensis* as a specific, safe, and effective tool for insect pest control. J Microbiol Biotechnol 17:547-549.

Salame L, Glazer I, Chubinishvilli MT, Chkhubianishvilli T (2010) Genetic improvement of the desiccation tolerance and host seeking ability of the entomopathogenic nematode *Steinemema feltiae*. Phytoparasitica 38:359-368.

Sandhu SK, Jagdale GB, Hogenhout SA, Grewal PS (2006) Comparative analysis of the expressed genome of the infective juvenile entomopathogenic nematode, Heterorhabditis bacteriophora. Molecular Biochemistry and Parasitology 145:239-244.

Sauphanor B, Berling M, Toubon J-F, Reyes M, Delnatte J, Allemoz P (2006) Carpocapse des pommes. Cas de resistance auvirus de la granulose en vergers biologique. Phytoma-La Defense Des Vegetaux 590:24-27.

Schnepf HE, Narva KE, Evans SL (2007) Modified chimeric Cry35proteins. US Patent 20077309785

Shapiro-Ilan DI, Gouge DH, Koppenhofer AM (2002) Factors affecting commercial success: Case studies in cotton, turf and citrus. In: Gaugler R, editor. Entomopathogenic Nematology. CAB International, New York 333-356pp.

Shelton AM, Romeis J, Kennedy GG (2008) IPM and GM, insect-protected plants: thoughts for the future. In: Romeis J, Shelton AM, Kennedy GG, editors. Genetically Modified Crops within IPM Programs. Springer, New York 419-429pp.

Shelton AM, Wang P, Zhao J-Z, Roush RT (2007) Resistance to insect pathogens and strategies to manage resistance: an update. In: LA Lacey, HK Kaya (eds), Field Manual of Techniques in Invertebrate Pathology. Springer, Dordrecht, 793-811pp.

Shim HJ, Choi JY, Wang Y, Tao XY, Liu Q, RohJY, Kim JS, Kim WJ, Woo SD, Jin BN, Je YH (2013) NeuroBactrus, a novel, highly effective, and environmentally friendly recombinant baculovirus insecticide. Appl Environ Microbiol 79(1):141-149.

Shukla N, Awasthi RP, Rawat L, Kumar J (2012). Biochemical and physiological responses of rice (*Oryza sativa* L.) as influenced by *Trichoderma harzianum* under drought stress. Plant Physiol Biochem 54:78–88.

Shukla N, Singh EANA, Kabadwa, BC, Sharma R, Kumar J (2019) Present status and future prospects of bio-agents in agriculture. Int J Curr Microbiol App Sci 8(4):2138-2153.

Singh HB, Singh A, Sarma BK, Upadhyay DN (2014). *Trichoderma viride* 2% WP (Strain No. BHU–2953) formulation suppresses tomato wilt caused by *Fusarium oxysporum* f. sp. *lycopersici* and chilli damping–off caused by *Pythium aphanidermatum* effectively under different agroclimatic conditions. Int J Agri Environ Biotechnol 7:313-320

Soberon M, Pardo-Lopez L, Lopez I, Gomez I, Tabashnik BE, Bravo A (2007) Engineering modified Bt toxins to counter insect resistance. Science 318:1640-1642.

St. Leger RJ, Joshi L, Bidochka MJ, Roberts DW (1996) Construction of an improved mycoinsecticide overexpressing a toxic protease. Proceedings of National Academy of Sciences USA, 93:6349-6354.

St. Leger RJ, Wang C (2010) Genetic engineering of fungal biocontrol agents to achieve greater efficacy against insect pests. Appl Microbiol Biotechnol 85:901-907.

Stewart LM, Hirst M, Ferber ML, Merryweather AT, Cayley PJ, Possee RD (1991) Construction of an improved baculovirus insecticide containing an insect-specific toxin gene. Nature 352:85-88.

Sun X, Chen X, Zhang Z, Wang H, Bianchi FJAA, Peng H, Vlak JM, Hu Z (2002) Bollworm responses to release of genetically modified *Helicoverpa armigera* nucleopolyhedroviruses in cotton. J Invertebr Pathol 81:63-69.

Sun X, Wu D, Sun X, Jin L, Ma Y, Bonning BC, Peng H, Hu Z (2009) Impact of *Helicoverpa armigera* nucleopolyhedroviruses expressing a cathepsin L-like protease on target and nontarget insect species on cotton. Biol Control 49:77-83.

Tabashnik BE, Gassmann AJ, Crowder DW, Corriere Y (2008) Insect resistance to Bt crops: Evidence versus theory. Nature Biotechnol 26:199-202.

Tabashnik BE, Unnithan GC, Masson L, Crowder DW, Li X, Corriere Y (2009) Asymmetrical cross-resistance between *Bacillus thuringiensis* toxins Cry1Ac and Cry2Ab in pink bollworm. Proc National Acad Sci USA 106:11889-11894

Tabashnik BE, Van Rensburg JBJ, Carriere Y (2009) Field evolved insect resistance to *Bt* crops: definition, theory, and data. J Econ Entomol 102:2011-25.

Tang XX, Sun XL, Pu GQ, Wang WB, Zhang CX Zhu J (2011). Expression of a neurotoxin gene improves the insecticidal activity of *Spodoptera litura* nucleopolyhedrovirus (SpltNPV). Virus Res 159(1):51-56.

Thakore Y. (2006) The biopesticide market for global use. Ind Biotechnol 2:194-208.

Tomalski MD, Miller LK (1991) Insect paralysis by baculovirus-mediated expression of a mite neurotoxin gene. Nature 352:82-85.

Wang CS, St. Leger RJ (2007) A scorpion neurotoxin increases the potency of a fungal insecticide. Nature Biotechnol 25:1455-56.

Wilson K, Cotter SC, Reeson AF, Pell JK (2001) Melanism and disease resistance in insects. Ecological Lett 4:637-49.

Wilson K, Yhomas MB, Blanford S, Doggett M, Simpson SJ, Moore SL (2002) Coping with crowds: density-dependent disease resistance in desert locusts. Proceedings of National Academy of Sciences USA 99:5471-5475.

Wilson KR, O'Reilly DR, Hails RS, Cory JS (2000) Age-related effects of the *Autographa californica* multiple nucleopolyhedrovirus egt gene in the cabbage looper (*Trichoplusia ni*). Biol Control 19:57-63.

Zhao JZ, Cao J, Li Y, Collins HL, Roush RT, Earle ED, et al (2003) Transgenic plants expressing two *Bacillus thuringiensis* toxins delay insect resistance evolution. Nature Biotechnol 21:1493-1497.

Zlotkin E, Fowler E, Gurevitz M, Adams ME (1993) Depressant insect selective neurotoxins from scorpion venom: chemistry, action, and gene cloning. Arch Insect Biochem Physiol 22:55-73.

Zwart MP, van der Werf W, Georgievska L, van Oers M, Vlak JM, Cory JS (2010) Mixed genotype infections of *Trichoplusia ni* larvae with *Autographa californica* multicapsid nucleopolyhedrovirus: Speed of action and persistence of a recombinant in serial passage. Biol Control 52:77-83.

9

Biopesticide Commercialization World-wide Regulation, Policies for Registration and Use of Biopesticides

Abstract

Biopesticides have become viable substitutes for man-made chemical pesticides in recent years. They are less expensive and don not endanger agro-ecosystems. Because of this, their demand and production are rising globally as well. In-depth examination reveals that there is no consistent regulatory approach that can streamline their regulation and registration procedure because the laws and regulations governing their usage and development differ from one country to the next. In spite of various effort by several international organisations like the Organization for Economic and Co-operative Development (OECD), International Organization for Biological Control (IOBC), and European and Mediterranean Plant Protection Organization (EPPO), some flexibility in biopesticide regulation has been offered, it still falls short of chemical pesticides, which have a strong market and well-established, non-overlapping legislation. World-wide regulation policies on biopesticide commercialization including registration and field use, limitations in regulations and modified regulations required are discussed in brief to understand the growth of biopesticides across the world.

Keywords: Biopesticide, Commercialization, Regulation policies Registration, Use

Introduction

The single piece of legislation under the Indian Government that regulates the import, manufacturing, sale, transportation, distribution, and use of all varieties of insecticides, including biopesticides, is the Insecticide Act (1968). Various parameters like shelf life, cross-contamination, moisture content, and packaging are significant factors that must be addressed before a biopesticide is registered. According to Organization for Economic Co-operation and Development's (OECD) recommendations, CIB simplified the protocols and listed the infrastructure needs for manufacturing of biopesticides. Information required to generate toxicity calls a strenuous effort. The demand for meteorological data, though, adds load on manufacturers and suppressed them from growing businesses. For instance, isolated microorganisms from one agroclimatic zone, possessing biocontrol property may or may not result the same findings in another agroclimatic zone. According to Rabindra (2005), Keswani et al. (2016), and available at http://ppqs.gov.in/divisions/cib-rc/guidelines, new biopesticides should go through provisional/temporary registration under either 9(3B) or 9(3) section of the Insecticide Act 1968 by providing information on moisture content, shelf life, commodity potency with reference to LC_{50}, toxicity, secondary non-pathogenic microbial.

Regulation Policies of Biopesticide Registration

Generally speaking, the organisms chosen for insect management are effective only against the target insect. Therefore, it is assumed that there is a minimal chance of hurting non-target creatures, such as people. Before authorizing the widespread use of biopesticides, it is required to conduct certain standardized safety tests that will support the presumption and provide evidence of their efficacy. As a result, the Food and Agriculture Organization of the United Nations has established criteria, accordingly several countries have also developed their own guidelines for licensing biopesticides (Kulshrestha 2004). Perhaps the most difficult aspect of biopesticides is their registration. The number of registered biopesticide products has increased recently, but this number could increase further if the registration process is standardised globally. There are many different authorities and laws emerging to control biopesticides, but very little latitude is offered. The laws observed in various nations and continents around the world are outlined in this section.

China

To regulate pesticide use and manufacturing in China, the Regulation on Pesticide Administration law was enacted in 1997. The law requires biopesticides to be registered before they can be sold (Kabaluk et al. 2010). Among other ministries, the Chinese Ministry of Agriculture (MOA) is authority of pesticide registration, manufacture, and commercial administration (Fang 2014). The Ministry of Agriculture's Institute for the Control of Agrochemicals (ICAMA 2008) is the apex regulatory body on monitoring the registration of pesticides, including biopesticides. The General Administration of Quality Supervision,

Inspection and Quarantine of the People's Republic of China only permits registered and approved businesses to submit applications for the registration of pesticides (Kabaluk et al. 2010). Through financial support for insect control in forests, the Chinese Ministry of Forestry promotes the use of biopesticide The good agricultural practices was inculcated among farmers by encouraging the use of biopesticides.

India

Government of India made many changes in regulations and laws to promote biopesticide manufacturers for registration. The Integrated Pest Management (IPM) initiative was overseen by the National Agricultural Technology Project (NATP) from 1998 to 2005, and the National Farmer Policy (2007) also supported the use of biopesticides in agriculture. By streamlining the licencing and regulating process for biopesticides, the Insecticide Act (1968) encouraged increased development and application of biopesticides. Under this act, the Central Insecticides Board (CIB) and the Registration Committee (RC) both functioned as extremely powerful entities for biopesticide regulation (www.cibrc.nic.in) (Kabaluk et al. 2010). The Apex Advisory Committee, or CIB, is composed of professionals from all relevant areas and fields. The CIB has simplified the criteria and data requirements for registration as well as the minimal infrastructure needs for the manufacture of biopesticides based on the OECD recommendations (NAAS 2013). After carefully examining and confirming claims on their bio-efficacy and safety for both humans and animals, the RC issues registrations. A key factor in the promotion of biopesticides is the National Agricultural Research System in which many ICAR institutes and State Agricultural Universities are involved (www.icar.org.in) (Rabindra 2005).

Africa

In order to create systems for the registration and regulation of biopesticides in the control of pests and diseases, some African nations adopt a variety of standards. Some African nations are taking the initiative to build their capabilities to control microbial pesticides. A regional inventory of the regulatory environments was conducted in 2012 by six country representatives from the West African region, including Kenya, Uganda, Ethiopia, Tanzania, Nigeria, and Ghana as part of the commercial Products (COMPRO II) project, which is run by the International Institute of Tropical Agriculture (IITA). The project's goal is to make biopesticides and biofertilizers more strictly regulated (Simiyu et al. 2013).

South Africa

The use, sale, and registration of biological control agents are governed by laws and regulations in South Africa. In accordance with Act 36 of 1947, the Department of Agriculture, Forestry, and Fisheries (DAFF) (www.daff.gov.za) regulates the registration of biological medicines (DAFF 2010).

European Union

In terms of the use and production of biopesticides, it is the second-largest continent. Microorganisms, plants, and pheromones were all governed under the EU's 1991 Directive 91/414/EEC, which was initially designed for chemical pesticides (Regnault-Roger et al. 2012). While new plant protection legislation was added in the EU in 2009, the following four pieces of legislation are also included: (1) Regulation (EC) No 1107/2009, (2) Directive 2009/128/EC, (3) Directive 2009/127/EC, and (4) Regulation (EC) No 1185/2009. The Directive 91/414 was amended by 2001/36/EC (EC 2001) and 2005/25/EC (EC 2005) to add the specific requirements for microorganisms. As of 2011, all member states must abide by the new Regulation (EC) No. 1107/2009, which takes the place of Directive 91/414/EEC (Meeussen 2012). The registration of biopesticides in EU nations appears to be more challenging than elsewhere in the world because the dossier must be submitted along with results of environmental and toxicological testing, as well as an efficacy assessment. According to Regulation (EC) No. 1107/2009, product registrations are handled by three zones based on geographic and climatic factors (Hauschild 2012). Denmark, Estonia, Latvia, Lithuania, Finland, and Sweden are in Zone A (North); Belgium, Czech Republic, Germany, Ireland, Luxembourg, Hungary, the Netherlands, Austria, Poland, Romania, Slovenia, Slovakia, and the UK are in Zone B (Central); and Bulgaria, Spain, Greece, France, Italy, Cyprus, Malta, and Portugal are in Zone C (South). Plant protection product (PPP) applicants must submit their registration dossier to a "Zonal reporters member state" (zRMS), which reviews the dossier. Regulation (EC) No. 283/2013, which was recently adopted, implements Regulation (EC) No. 1107/2009 for establishing data-related concerns (EC 2013).

USSR (formerly)

The Russian Agricultural Control regulates the state registration of microbiological pesticides in Russia (RAC). In addition to managing pesticide registration, RAC oversees pesticide usage, manufacture, sale, transportation, storage, disposal, advertising, import, and export (Kabaluk et al. 2010). The Russian Agricultural Academy (RAN), which houses the All-Russian Institute for Plant Protection (VIZR) in St. Petersburg, is involved in the registration procedure as well as research and development of biopesticides (Kabaluk et al. 2010).

United Kingdom

The Chemicals Regulatory Directorate (CRD)/Pesticide Safety Directorate (PSD) (http://www.hse.gov.uk/pesticides/) is the primary regulatory authority in the UK in charge of plant protection products, including biopesticides. Pesticides, biocides, detergents, and other chemicals are regulated by the CRD, a new Directorate of the Health and Safety Executive (HSE), in accordance with the Registration, Evaluation, Authorization and Restriction of Chemicals Act (REACH). Agricultural pesticide registration is handled by the Department of Environment, Food, and Rural Affairs (DEFRA) entity known as PSD (DEFRA 2006). The UK regulatory framework was created on a chemical pesticide model, which might have prevented the commercialization of biopesticides (ACP 2004). The biopesticide scheme was created in 2003 as a significant initiative, and its primary goal was to increase the production of biopesticides (http://www.pesticides.gov.uk/environment.asp). In order to register and regulate biopesticides, this approach introduced the position of "biopesticide champion" in 2006 (Chandler et al. 2011).

USA

A sizeable component of the worldwide biopesticide market is in the United States. According to USEPA (2010), the EPA in the USA has a comprehensive and complicated regulatory system for the registration and regulation of biopesticides, and this system has registration requirements that are different from those of other regulatory systems (Harman et al. 2010; Chandler et al. 2011). The Office of Chemical Safety and Pollution Prevention (OCSPP) and the Office of Pesticide Programs (OPP) are in charge of regulating biopesticides, and OPP is divided into three divisions that are involved in pesticide registration: the Antimicrobial Division, the Registration Division, and the Biopesticides and Pollution Prevention Division (BPPD) (Matthews 2014). EPA typically mandates the use of biopesticides since they pose fewer dangers than chemical pesticides.

The Insecticides Act of 1968, which was endorsed by the pesticide registration committee in India, set the rules for the registration of biopesticides. The research and commercialization of pest control solutions involves a number of stakeholders, including scientists, regulators, marketers, and end users. Although some members of this chain are frequently involved from the very beginning of the development process, there are still many problems to be solved. For example, marketers may frequently disagree with regulators and scientists, leaving end users perplexed about alleged flaws in the finished product (Damalas and Koutroubas 2018; Satapathy 2018).

Biopesticide Registration Protocol in India

Regulations for biopesticide registration and further marketing were framed during the 357th Meeting of CIB&RC held on 10th August, 2015. Important regulations are hereunder.

1. The earlier registrants of the strain/inventor of the strain has to deposit one sample containing at least one kg of product/formulation to the Secretary, CIBRC that should be subject to 16 SR-DNA/Gene code sequencing/finger printing for creating data bank of all the strains by ICAR-NBAIM, Mau.
2. Registration of already registered strains of biopesticides:
   a. Requirement of the data/information to be submitted for getting permanent registration under section 9(3) /9(3B)
   b. Form-I duly filled in along with requisite registration fee of Rs. 100 as per existing requirement.
   c. Already approved Label leaflets of the product/strain
   d. Testimonial/documents about the company as per existing requirement.
   e. Undertaking about the strain from the inventor of the strain or first registrant or subsequent registrant of the strain or the applicant.
   f. One sample (minimum one kg) for pre-registration verification (PRV) through Central Insecticides Laboratory
   g. Another sample (minimum one kg) for pre-registration verification (PRV) of Gene code sequencing/16 SR-DNA/finger printing along with a demand draft (as per invoice obtained as testing fee from NBAIM, Mau) in favour of NBAIM, Mau as testing fee for Gene code sequencing/16 SR-DNA/finger printing.
3. Registration of new strain of the biopesticides:
   a. The applicants for registration of new strain has to submit all the data as per existing guidelines for registration under section 9(3)/9(3B) for all the disciplines. Two samples have to be submitted to the Sectt. of CIB&RC; one for pre-registration verification (PRV) from Central Insecticides Laboratory as per product specification requirement & another sample to be used for pre-registration verification (PRV) of Gene code sequencing/16 SR-DNA/finger printing along with a demand draft (as per invoice obtained as testing fee from NBAIM,

Mau) in favour of NBAIM, Mau as testing fee for Gene code sequencing/16 SR-DNA/finger printing

b. Minimum infrastructure required for production and registration of biopesticides:
c. Verification of the infrastructure and technical competency of the applicants already registered under section 9(3B) and applying for registration u/s 9(3) and/9(3B) extension has to be conducted by a team constituted by the Secretary (CIB&RC) for the purpose.
d. Minimum CFU count and nominal concentration strength of the formulation to be continued as per existing guidelines
e. Verification of shelf life of strain and verification of product
f. Submission of photographs for veracity of research, test and trails

Policies on Biopesticide Use

The National Agriculture Policy of India from 2000 placed a strong emphasis on farmers receiving timely and appropriate supplies of agricultural inputs, including biopesticides. In accordance with the "Zero Budget Natural Farming" (ZBNF) initiative put forth by the Food and Agriculture Organization of the United Nations (FAO), which promotes the use of locally obtainable natural fertilisers and biopesticides as well as farmer-owned seeds for organic farming, the Government of India has taken the necessary coordinated action (https://www.fao.org/agroecology/detail/en/c/443712/).

The marketing of biopesticides to farmers is the responsibility of the Ministry of Agriculture and Farmers Welfare and the Department of Biotechnology (DBT), in addition to the Central Integrated Pest Management Centre (CIPMC), Faridabad, the National Centre for IPM (NCIPM) under the Indian Agricultural Research Council, and the Directorate of Biological Control (Alam 1994). The Department of Biotechnology (DBT), in addition to the aforementioned regulatory bodies, funds research into the development of biopesticides (Sinha and Biswas 2008). Both the National Accreditation Board (NBA) and the National Agricultural Research System (NARS) conduct quality control testing on biopesticides and train state agricultural departments in these techniques.

The registration procedure for biopesticide products seems to be impeding their commercialisation. In order to enable quick registration of biopesticide products based on justifiable standards, regulatory agencies should encourage the use of safer technology in the creation of commercial products. Additionally, the regulatory framework should support the growth of small and medium-sized biopesticide companies, enabling them to provide consumers high-quality products and giving growers trustworthy tools for the cost-effective control of pests. Data requirements for biological goods are frequently derived from those for synthetic chemical products. However, risk assessment for biopesticides ought to be based on pertinent scientific knowledge rather than synthetic chemical criteria. In order to reflect the nature of the various categories of biopesticide active ingredients, it is required to modify the standards. Data standards and instructions for biopesticides are now accurately modified (Isman 2014). The major problem for the biocontrol sector seems to be the length of submission procedures at both the EU and Member State levels. In order for new products to succeed on the market, faster processes and the enforcement of deadlines are essential. The high expense associated with registering new medicines is another barrier to the commercialization of novel products (Pavela 2014).

Legislation that prohibit the use of conventional pesticides like dicrotophos, azinphos ethyl, and ammonium sulphate, among others, that have been passed by governments of countries like India, Germany, and other European nations. These laws are projected to help the biopesticides market grow as a result of the extraction of pesticides from natural resources such as animals, plants, microbes, and particular minerals. It was predicted that the ban on a certain class of chemical pesticides might affect crop exports from India to other countries, particularly Europe. For instance, the Agricultural and Processed Food Products Export Development Authority between the United States and India has decided to end their collaboration as of 2020 APEDA. All organic businesses in India that seek to export to the US after July 2022 needed a certificate provided by a USDA-accredited certifier in order to export the generated crops. This requirement took effect after an 18-month transition period. For instance, the export amount of rice declined from 2018 to 2019 as a result of limits on specific chemicals, and it was projected that the use of biopesticides might assist rice growers increase the export volume in 2020. The usage of biopesticides was therefore expected to rise during the anticipated period as a result of rules regarding the use of chemicals for crop protection. (https://www.researchandmarkets.com/reports/5175605/india-biopesticides-market-growth-trends-and#rela2-5214644).

Limitations in Regulations of Biopesticide Registration

The primary issue with biopesticide regulation, which is a systemic one, is that it is based on the models used for traditional chemical pesticides (Greaves 2009). In the EU system, regulatory failure, according to Chandler et al. (2008), results from the use of an ineffective synthetic pesticide paradigm and a lack of regulatory innovation. In addition, the evaluation of biopesticides and their registration for commercial use are also drawn-out processes. The sector complains that the current registration period is expensive and time-consuming, notably for microbial biological control, and that the EU system takes a long time to process registrations (Bailey et al. 2010). As an illustration, the average time required in the EU was 75 months as opposed to 28 months in the USA (Hokkanen and Menzler-Hokkanen 2008). The US approach is flexible, and it occasionally invites applicants to pre-submission meetings where the applicant is informed on which investigations are required, based on available literature and preliminary data (Mubyana-Jhon and Taylor 2015).

Asia's biopesticide production system is underdeveloped and underutilised as a result of a number of institutional, social, and technical barriers that prevent the commercial production of innovative biopesticides (NAAS 2013). There is a difficulty with quality control in developing nations like Asia and Africa, which makes it difficult for farmers to have confidence in their products. Only an effective regulatory structure will be able to remedy this issue. Even though India is producing and using biopesticides, the increase is still lagging behind that of chemical pesticides. In a study, Rabindra (2005) projected that less than 10% of the identified need is being met by existing production of microbial pesticides. The CIB has registered around 500 biopesticides, which are available on the Indian market. However, quality control is a significant problem for the majority of the products (NAAS 2013).

Even though data requirements are becoming more transparent and standardised for more effective regulatory procedures, Mensink and Scheepmaker (2007) contend that insufficient guidance on the evaluation and use of biological products prevents premarket evaluations of the environmental safety from being carried out. It is difficult to establish an evaluation method that is equally fair to both biopesticides and chemical pesticides since regulatory authorities are aware that biopesticides are fundamentally different from chemical pesticides and should not be evaluated with the same standards of safety and efficacy (Bailey et al. 2010). One issue is that regulatory mechanisms only evaluate individual items, although the nature of microbial pesticides is extremely complicated and varied (Hubbard et al. 2014). In addition to providing guidelines, Ravensberg (2011) provided advice on how to compile a dossier and what sources to consult in order to better comprehend the exact requirements of the authorities. A data requirement is rather widespread in the USA and Canada, however the EPA does not call for a comprehensive dossier of efficacy and phytotoxicity data, while PMRA does.

The regulation of biopesticides with diverse modes of action is another complicated problem. For instance, *Trichoderma* species that are utilised as biopesticides against soil borne plant pathogenic fungi can parasitize such fungi in the soil; they can also create antibiotics (Ghisalberti and Sivasithamparam 1991; Vey et al. 2001) and enzymes that break down fungal cell walls (Bech et al. 2015). *Trichoderma* compete with soil borne pathogens for carbon, nitrogen, and other resources (Limon and Codon 2004). They can also encourage plant development by producing chemicals that are similar to auxin (Vinale et al. 2008; Nega 2014). Some *Trichoderma* products have been marketed as plant growth promoters rather than plant pesticides (Nega 2014), which has allowed them to avoid regulatory review of their effectiveness and safety (Bailey et al. 2010). Pseudomonas is in a same situation. Fluorescent Pseudomonas can be employed for both biocontrol and encouraging plant development (Negi et al. 2005; Mehnaz 2013; Tewari and Arora 2014). There aren't any specific regulatory controls in place to prevent this, though. To effectively use biopesticides, there are a number of technological and regulatory gaps that must be filled in order to reduce the use of chemical pesticides and to advance the use of biopesticides (Kumar 2015).

Interventions in Regulations of Biopesticide

In order to increase the production of agrobiologicals on a global scale, innovation in the current biopesticide control framework is essential (Arora et al. 2012). Currently, the regulatory environment differs by country; some have developed systems, some are making progress in their regulatory frameworks, and a few do not have adequate rules for biopesticides (Simiyu et al. 2013). In order to effectively control pests, the regulatory structure in place should be environmentally benign, scientifically sound, and technologically advanced (Greaves 2009). The issue of why biopesticides aren't utilised more frequently could be resolved by cutting registration fees and doing away with effectiveness requirements (Greaves 2009). Another sensible strategy for improving biopesticide regulation is for nations to enact laws on a worldwide scale by holding conferences, workshops, and meetings to raise the status of biopesticides (Mishra et al. 2015).

Exogenous pressure, such as government action, and endogenous pressure, such as pressure within regulatory organisations, are some variables that may encourage the necessary regulatory improvements (Greaves 2009). There is a need for guidelines to encourage the collaboration of businesses and research institutes because several institutions have conducted some preliminary research about the industrialization of biopesticides and institutional changes may be significant; however, no systematic reports have yet been published

(Leng et al. 2014). The innovative strategy for the manufacturing and marketing of biopesticides depends heavily on the global harmonisation of biopesticide regulatory rules, and the OECD is crucial for this harmonisation at the global level (Holm et al. 2005). The World Health Organization (WHO) and the OECD have an impact on pesticide control, and their participation is crucial (NAAS 2013). In order to assist its member nations in harmonising the methods and procedures used to analyse biological pesticides, the OECD project on biopesticides was launched in 1999 (Sigman 2005).

More than 70 emerging and transition economies have working links with the OECD, which now comprises 34 member countries (http://www.oecd.org/chemicalsafety/pesticides-biocides/). The OECD's working group on pesticides is made up of the I Registration Steering Group (RSG), (ii) Risk Reduction Steering Group (RRSG), and (iii) Biopesticides Steering Group (BPSG). Through the creation of working documents and guidance, the BPSG has made significant strides toward harmonisation and work sharing (Richards and Kearns 1997). The OECD group's headquarters are in Paris, France, and they work closely with EU governments to carefully examine the risks that biopesticides pose to people and the environment (http://www.biopesticideindustry alliance.org/).

The OECD, Food and Agriculture Organization (FAO), and EU have all focused their emphasis on pesticide control globally, in general and in specific (Greaves and Grant 2011; FAO 2012). An intergovernmental organisation in Paris called EPPO is financed by contributions from its member nations (www.eppo.int). The International Organization for the Control of Noxious Animals and Plants (IOBC) examined the rapid global spread of the use of microbial pesticides and improvements in their regulatory systems in 2010 (IOBC 2010). Various organisations, including the OECD, North American and European governments, have made significant strides toward promoting harmonisation for biopesticide legislation and facility developments for work sharing between governments (AGBR 2015). As the most significant international organisations for biopesticide regulation and innovation, the BPSG of the OECD, the FAO, the European Commission (EC), the IOBC, the EPPO, the North American Plant Protection Organization (NAPPO), and NAFTA have all worked together (AGBR 2015).

Conclusions

Worldwide, the commercialization of biopesticides is expanding quickly, however the increase is not proceeding as anticipated due to a lack of appropriate laws and other restrictions. Effective regulation can also stop fake biopesticides from being sold. The regulation of biopesticides is a barrier to their manufacturing and commercialization. The aforementioned explanation makes clear that different countries have different regulation criteria. It is suggested that a common regulatory system be developed to overcome this obstacle. In order to examine the potential dangers related to microbial biopesticides, it is also necessary to increase communication and information exchange between regulatory agencies, scientists, and enterprises. Guidelines for assessing the efficacy, quality, and field testing of biopesticides also need to be updated because they are frequently carried out by non-experts, particularly in poor nations. Establishing regulatory organisations is necessary to ensure quick registration of biopesticide products with justifiable regulations and open processes, as well as to support the adoption of new, safer technology in the creation of commercial products. The standard for regulating should be the same across all nations and should be based on the nature of agro-biologicals rather than chemical pesticides.

References

ACP (2004) Advisory Committee on Pesticides. Final report of the sub-group of the advisory committee on pesticides on: alternative to conventional pest control techniques in the UK: a scoping study of the potential for their wider use. Advisory Committee on Pesticides,York. http://www/pesticides.gov.uk/uploadedfiles/WebAssests/ACP/ACP_alternatives_web_subgrp_report.pdf.

AGBR (2015) Agrow Global Biopesticide Regulations. Commodity analysis. Available at www.agra-net.com.

Alam G (1995) Biotechnology and sustainable agriculture: Lessons from India. OECD development centre, working paper no. 103, Sustainable Development: Environment, Resource Use, Technology and Trade.

Arora NK, Tewari S, Singh S, Lal N, Maheshwari DK (2012) PGPR for protection of plant health under saline conditions, 239-258pp. In: DK Maheshwari (ed), Bacteria in agrobiology: stress management, Springer, Berlin.

Bailey KL, Boyetechko SM, Langle T (2012) social and economic drivers shaping the future of biological control: a Canadian perspective on the factors affecting the development and use of microbial biopesticides. Biol Control 52:221-229.

Bech I, Busk PK, Lange L (2015) Cell wall degrading enzymes in *Trichoderma asperellum* grown on wheat bran. Fungal genome Biol 4:116.

Chandler D, Bailey AS, Tatchell GM, Davidson G, Greaves J, Grant WP (2011) The development, regulaltion and use of biopesticides for integrated pest management. Philos Trans R Sco Lond H: Biol Sci 366(1573):1987-1998.

Chandler D, Davidson G, Grant WP, Greaves J, Tatchell GM (2008) Microbial biopesticides for integrated crop management: an assessment of environmental and regulatory sustainability. Trends Food Sci Technol 19:275-283.

DAFF (2010) Department of Agriculture, Forestry and Fisheries. Act No. 36 of 1947, Guidelines on the data required for registration of biological/biopesticides remedies in South Africa, Republic of South Africa.

Damalas CA, Koutroubas SD (2018) Current status and recent developments in biopesticide use. Agriculture 8(13). doi:10.3390/agriculture8010013.

DEFRA (2006) Department for Environment, Food and Rural Affairs. The royal commission on environmental pollution report on crop spraying and the health of residents and bystanders-governmentresponse.http://www.defra,gov.uk/environment/rcep/index.htm.

EC (2013) European Commission. No. 283/2013 setting out the data requirements for active substances, in accordance with regulation (EC) No. 1107/2009 of the European parliament and of the council concerning the placing of plant protection products on the market. Off Eur Union L 93:1-84

FAO (2012) Food and Agriculture Organization. Guidance for harmonizing pesticide regulatory management in Southeast Asia RAP publication 2012/13. Food and Agriculture Organization of the United Nations, Regional Office for Asia and the Pacific, Bangkok, p 480.

Ghisalberti EL, Sivasithamparam K (1991) antifungal antibiotics produced by *Trichoderma* spp. Soil Biol Biochem 23(11):1011-1020.

Greaves J (2009) Biopesticides, regulatory innovation and the regulatory state. Public Policy Adm 24(3):245-264.

Greaves J, Grant WP (2011) The development, regulation and use of biopesticides for integrated pest management. Phil Trans R Soc B 366:1987-1998.

Harman GE, Obregon MA, Samuels GJ, Lorrito M (2010) Changing models for commercialization and implementation of biocontrol in the developing and developed world. Plant Dis Int J Appl Plant Pathol 94(8):928-993.

Hauschild R (2012) Safety and regulation of microbial pest control agents and microbial plant growth promoters: Introduction and overview, 67-71pp. In: I. Sundh, A Wilcks, MS Goettel (eds), Beneficial microorganism in agriculture, food and the environment: safety assessment and regulation. CAB International, Wallingford.

Hokkanen H, Menzler-Hokkanen (2008) Deliverable 24: cost, trade-off and benefit analysis, In: Regulation of biological control agents: final activity report. Christain-albrechts-University, kiel.

Holm RE, Baron JJ, Kunkel DL (2005) World horticulture in crisis, 31-40pp. In: The BCPC international congress proceedings, British Crop Protection Council, Alton.

ICAMA (2008) Institute for the Control of Agrochemicals, Ministry of Agriculture. Pesticide manual, the institute for the control of agrochemicals. Ministry of Agriculture, China. www.agrolex.com.cn.

IOBC (2010) International Organization for Biological Control. Global newsletter; Issue 88. http://www.iobc-wprs.org/pub/.

Isman MB, Grieneisen ML (2014) Botanical insecticide research: Many publications, limited useful data. Trends Plant Sci 19:140-145.

Kabaluk JT, Antonet MS, Mark SG, Stephanie GW (2010) The use and regulation of microbial pesticides in representative jurisdictions worldwide. IOBC Global.

Keswani C, Sarma B, Singh H (2016) Synthesis of Policy Support, Quality Control, and Regulatory Management of Biopesticides in Sustainable Agriculture. 10.1007/978-981-10-2576-1_1.

Kulshrestha S (2004) Status of regulatory for biopesticides in India, 67-71pp. In: Kaushik N (ed), Biopesticides for sustainable agriculture: prospects and constraints, Teri Press, The Energy and Resource Institute, New Delhi.

Kumar S (2015) Biopesticides: a need for food and environmental safety. J Biofertil Biopestici 3:e127.

Leng P, Zhang Z, Pan G, Zhao M (2014) Applications and development trends in biopesticide. Aft J Biotechnol 10(8):19864-19873.

Limon MC, Codon AC (2004) Biocontrol mechanisms of *Trichoderma* strains. Int Microbiol 7:249-260.

Matthews KA (2014) Regulaltion of biopesticides by the environmental protection agency, 267-279pp. In: JR Coats (ed), Biopesticides: state of the art and future opportunities, vol 18, Symposium series, 1172, Sidley Austin, Washington, DC.

Meeussen J (2012) OECD guidelines and harmonization for microbial control agents, 308-321pp. In: I Sundh, A. Wilcks, MS Goettel (eds), Beneficial microorganism in agriculture, food and the environment: safety assessment and regulation. CAB International, Wallingford.

Mensink BJWG, Scheepmaker JWA (2007) how to evaluate the environmental safety of microbial plant protection products: a proposal. Biocontrol Sci Technol 17:3-20.

Mishra J, Tewari S, Singh S, Arora NK (2015) Biopesticides: Where we stand?, 37-75pp. In: NK Arora (ed), Plant microbes symbiosis: applied facets, Springer, New Delhi.

Mubyana-John T, Taylor J (2015) Nematophagous fungi: regulations and safety, 203-213pp. In: TH Askary, PRP Martinelli (eds) Biocontrol agents of phytonematodes, CABI, Wallingford.

NAAS (2013) National Academy of Agricultural Sciences. Biopesticides- quality assurance. http://naasindia.org/Policy%20Papers/Policy%2062.pdf.

Nega A (2014) Review on concepts in biological control of plant pathogens. J Biol Agric Healthc 4(27):33-54.

Negi YK, Garg SK, Kumar J (2005) Cold-tolerat fluorescent *Pseudomonas iisolates* from Garhwal Himalayas as potential plant growth promoting and biocontrol agents in pea. Curr Sci 89:2151-2156.

Pavela R, Benelli G (2016) Essential oils as ecofriendly biopesticides?. Challenges and constraints. Trends Plants Sci 21(12):1000-1007.

Rabindra RJ (2005) Current status of production and use of microbial pesticides in India and the way forward, 1-12pp. In: RJ Rabindra, SS Hussaini, B Ramanujam (eds), Microbial biopesticide formulations and applications Technical Document no. 55. Project Directorate of Biological Control.

Ravensberg WJ (2011) A roadmap to the successful development and commercialization of microbial pest control products for control of arthropods, Springer, Dordrecht 171-233pp.

Regunault-Roger C, Vincent C, Amason JT (2012) Essential oils in insect control: low-risk products in a high stakes world. Ann Rev Entomol 57:405-424.

Richards J, Kearns P (1997) Work on micro-organisms within the OECD's environmental health and safety programme. In: Evans HF (ed) Microbial insecticides: novelty or necessity? BCPC symposium proceedings no. 68. Coventry BCPC, Farnham, pp 223-226.

Satapathy SN (2018) Regulatory norms and quality control of bio-pesticides in India. Int J Curr Microbiol App Sci 7(11):3118-3122.

Sigman R (2005) OECD's vision for the regulation of crop protection products, 329-336pp. In: proceedings of the BCPC International Congress-Crop Sci & Technol, Glasgow.

Simiiyu NSW, Tarus D, Watiti J, Nang'ayo F (2013) Effective regulation of bio-fertilizers and bio-pesticides: a potential avenue to increase agricultural productivity, Policy series, COMPRO II, IITA.

Sinha B, Biswas I (2008) S&T for rural India and inclusive growth, potential of biopesticide in Indian agriculture vis-à-vis rural development. Ind Sci Technol. WWW.bistads.res.in/indiansnt2008/t6rural/t6rur17.htm.

Tweari S, Arora NK (2014) Multifunctional exoploysaccharides from *Pseudomonas aeruginosa* PF23 involved in plant growth stimulation biocontrol and stress amelioration in sunflower under saline conditions. Curr Microbiol 69(4):484-494.

USEPA (2010) United States Environmental Protection Agency. Prevention, pesticides and toxic substances, Series 885-microbial pesticides test guidelines. EPA, Washington, DC.

Vey A, Hogland RE, Butt TM (2001) Toxic metabolites of fungal biocontrol agents 311-346pp. In: TM Butt, C Jackson, N morgan (eds), Fungi as biocontrol agent: progress, problem and potential, CAB International, Bristol.

Vinale F, Sivasithamparam K, Ghisalberti EL, Marra R, Woo SL, Lorito M (2008) *Trichoderma* plant-pathogen interactions. Soil Biol Biochem 40:1-10.

Links

http://www.biopesticideindustry alliance.org/

https ://www.fao.org/agroecolog y/detail/en/c/443712/

https://www.researchandmarkets.com/reports/5175605/india-biopesticides-market-growth-trends-and#rela2-5214644

10

Promotion of Biopesticides in India Role of Government and Growers

Abstract

Biopesticides are cutting-edge crop protection agents that shield crops from a wide range of pests and pathogens in an environmentally responsible way. They outperform synthetic pesticides in a wide range of ways, including target specificity, reduced toxicity, and biodegradability. Despite this, they are underrepresented in the crop protection industry, accounting for only 3.5 percent of the worldwide pesticides market. Biopesticides have a lot to offer for the development of sustainable agriculture, despite their slow adoption in the commercial pesticide industry. Understanding the main obstacles and constraints that affect the market for biopesticides can help in the development of innovative approaches including improving delivery systems, selecting new and improved strains, and preparing farmers and other stakeholders to deal with issues.

Keywords: Biopesticide, Promotion, Initiatives, Consumer awareness

Introduction

Pest and pathogen incidence is a natural occurrence that frequently goes unreported. However, they become a concern when their spectra expand and cause significant losses. Chemical pesticides are now commonly used in agricultural techniques to lessen the impact of such severe damages. Without a question, the use of chemical pesticides has put human health, ecological health, and sustainability at danger. Therefore, the use of biopesticides in pest management programmes has been recognised as a sustainable solution to free agriculture from the debt of disease occurrence and insect infestation. Increased organic farming areas and related efforts in India, such as SOM, NPOP, SMPMA, NMSA, PKVY, ZBNF, etc., are thought to support the market and use of biopesticides.

Government Initiatives

Due to the subsidy component/incentive on conventional pesticides, the current agro-industry is reticent to do research and produce biopesticides. However, due to restrictions on the broad use of chemical pesticides and the phasing out and banning of a few toxic substances, there has been an increased push in recent decades to develop biopesticides for commercial usage. The percentage share of biocontrol products is still considerably lower than that of chemicals, though. Policies such as entrepreneurial education, institutional finance availability, subsidies, insurance, and tax and duty exemption can all increase the production of biopesticides. Government support for the use of biopesticides and the designation of no-pesticide zones may help the situation for bioproducts. For example, the Sikkim Organic Mission (SOM), which converted about 75,000 hectares of agricultural land, is now India's first organic state as a result of more strictly enforcing the National Programme for Organic Production (NPOP) criteria connected to the organic mission. Examining the SOM model, it was discovered that in this situation, producers and authorities were urged to employ organic inputs while avoiding synthetic ones. The similar idea of becoming organic is also being tried in Kerala, Arunachal Pradesh, and Mizoram.

In order to advance the organic movement and lower chemical risk, the Ministry of Agriculture's Department of Agriculture & Cooperation introduced the Organic Farming Policy in 2005. The regulation recognised organic sources of nutrients such biofertilizers, organic manures, compost, and biocontrol agents as certified inputs for organic farming (biopesticides). The National Bank for Agriculture and Rural Development (NABARD) introduced the Strengthening and Modernizing Pest Management Approach in India (SMPMA) capital investment subsidy programme, which provided financial assistance for the establishment of bio-fertilizer/bio-pesticide units as a 25% subsidy up to a maximum of 4 million rupees.

The National Action Plan on Climate Change included the establishment of the National Mission for Sustainable Agriculture (NMSA), which dealt with "Sustainable Agriculture" issues (NAPCC). The third mission intervention of NMSA was related to pest management and aimed to promote biopesticide research, commercial manufacturing, and commercialization. The major objective was to develop new biopesticides and technology for disease prediction employing innovative botanical applications, sterile insect approaches, transgenic insects, semiochemicals, and endophytic microbial metabolites. Additionally, the "*Paramparagat Krishi Vikas Yojana*" (PKVY) and "Soil Health Management" (SHM) programmes have been launched to support organic farming by adopting organic communities through a cluster model and PGS certification (Reddy 2017). Farmers ensure their product is free of any synthetic chemicals, including fertilisers, pesticides, and hormones, under the self-regulatory PGS programme. A neighbourhood group of five or more organic farms supports the programme. The PGS Organic Council unifies the standards for production quality control and permits the use of its PGS label as a quality stamp on goods (https://www.pgsorganic.in).

In the last five years, the government has also taken the required steps to support the widespread use of biopesticides. The "Zero-Budget Farming" method, which has had considerable success in southern India, is already in use in a few other states across the country. Zero Budget Natural Farming (ZBNF), as the technique is known by the Food and Agriculture Organization of the United Nations (FAO), emphasises minimising the superfluous expenditure of agricultural inputs such the purchase of pricey seed, chemical fertilisers, and pesticides. Instead of such expensive machinery, it encourages the use of farmer-owned seeds, naturally occurring local fertilisers, and biopesticides for organic farming.

Consumer Awareness on Biopesticides

The inadequate usage of biopesticides is primarily the result of consumer and user ignorance (Arora et al. 2010). The word "biopesticides" is unfamiliar to many farmers, and others are unsure whether to use them instead of chemicals. Because of imprecise and unfavourable results, several people have stopped using biopesticides. However, since the production techniques utilised in such formulations do not meet the requirements specified by regulatory agencies, low-quality, non-registered products are particularly affected by this issue. It is imperative to stress that the proposed biopesticides should have reliable, repeatable, consistent, and focused activity (Mishra et al. 2015). The host range and circumstances under which the formulation will work should be clearly stated on the product. Knowing how farmers feel about biopesticides is important because they are the ultimate users of these products. This is because it helps determine the suggestions and needs for appropriate biological control measures in farming systems. However, there is a marked difference between small and large farmers in adopting biopesticides in practice.

Smaller farmers frequently ignore or disregard government initiatives and programmes pertaining to organic agriculture. Additionally, there are myths about biopesticide requirements such as higher costs, lesser yield, and other requirements. Furthermore, the illicit sale and usage of counterfeit goods is a serious problem that has caused farmers to lose faith in biopesticides; this calls for prompt government attention (FICCI 2015). By offering orientation and demonstration sessions where farmers may learn how to use quality products, private enterprises may also help to resolve this issue. For the same reason, farmer field schools (FFS), which offer field-based, location-specific instruction on biopesticides for the development of knowledge and confidence among the end-users, have been established in various states (Mohanty and Sahu 2019).

Despite being safer for the environment than synthetic pesticides, the use of biopesticides is far lower than that of synthetic pesticides due to lack of information, lack of trust, and unavailability in local markets. In order for governmental, non-governmental, and educational institutions to properly apply biopesticides, farmers must be educated about the advantages of doing so through on-farm training. The government should offer the biopesticides as a subsidy or for free when purchasing agricultural products to promote their use by farmers.

Conclusions

Although biopesticides have demonstrated their value in the sustainable management of pests and pathogens, they are currently a niche product in the crop protection market. The primary causes of the market for biopesticides still being in its infancy are farmers' lack of confidence as a result of their long-standing reliance on chemical pesticides for crop protection, their lack of awareness, poor government support, a lax regulatory system, inappropriate technologies, and a lack of knowledge. Governmental, non-governmental, corporate, and public institutions, as well as universities, must all take a holistic approach to meet these difficulties.

References

Arora NK, Khare E, Maheshwari DK (2010) Plant growthpromoting rhizobacteria: constraints in bioformulation, commercialization, and future strategies 97-116pp. In: D Maheshwari (ed) Plant growth and health promoting bacteria. Microbiology Monographs, vol 18, Springer, Heidelberg.

FICCI (2015) Study on sub-standard, spurious/counterfeit pesticides in India 2015-report. Federation of Indian Chambers of Commerce & Industry 96p.

Mishra J, Tewari S, Sing S, Arora NK (2015) Biopesticides: where we stand? 37-75pp. In: N Arora (ed) Plant microbes symbiosis: applied facets, Springer, New Delhi

Mohanty A, Sahu R (2019) Farmer Field School-building knowledge on the farm LEISA INDIA available online at https://leisaindia.org/farmer-field-school-building-knowledge-on-the-farm/

Reddy AA (2017) Impact study of paramparagat krishi vikas yojana. National Institute of Agricultural Extension Management (MANAGE), Hyderabad 210p.

Links

https://www.pgsorganic.in

11

Growth of Biopesticides Driving Force and Set-Back

Abstract

The toxicity and non-biodegradability of chemical pesticides have stoked the demand for more sustainable alternatives. In addition to this, the persistently rising demand for cost-effective pest control measures has increased the utilization of biopesticides across several countries. However, their use has remained low in certain under-developed nations but are expected to witness better growth in the coming years. India offers a wide range of options in terms of supplies for natural biological control organisms as well as natural plant-based insecticides because of its great biodiversity. The widely diverse indigenous tribes in India's rich traditional knowledge base may hold important hints for the development of more advanced and efficient biopesticide. The National Farmer Policy of 2007 aggressively encouraged the adoption of biopesticides to boost agricultural output while maintaining farmer and environmental health. Additionally, it states that biopesticides will receive the same funding and promotion as chemical pesticides. Biopesticides are yet to take off in a major way in India because of mixed constraints, despite their enormous market potential and the national and state initiatives to promote them as alternatives to chemical pesticides. This chapter seeks to examine the factors enabling growth in the market as well as those restraining its trajectory.

Keywords: Biopesticide, Adaptation, Driving force, Demerits

Introduction

Regulations should make it easier to utilise creative, long-lasting solutions, allowing for the selection of the most environmentally friendly pest management method. This can be accomplished by using expedited registration, priority registration, and a combination of comparative evaluation of pest control techniques and the substitution principle, which allows a natural pest control technique to take the place of a synthetic pesticide. More microbial biological control agents will be registered more quickly as a result of the modifications in registration procedures, which will logically lead to lower product costs (EC 2009; van Lenteren et al. 2018).

Another significant step toward making biological control more appealing and available to farmers is the development of a standardised process for the registration of microbial biological control agents that may be used locally or globally. Use of biopesticides is prompted by the removal of pesticides from the market as a result of observed health, non-target, and environmental effects, the emergence of new pests for which no pesticides are available, the development of resistance that reduces the effectiveness of pesticides, and all stimulate use of biopesticides (Urbaneja et al. 2012). Non-governmental organisations (NGOs) have had success switching from chemical to biological control in a number of instances by providing information on the impacts of pesticides on the environment and their illegitimate usage (Calvo et al. 2012).

A rise in the use of biological control has also been attributed to the development of new and improved biological control methods, improved and more stable formulations for microbial biological control agents and their use as seed treatments, more practical application techniques for invertebrate biological control agents (equipment to release biological control agents in crops, use of drones, etc.), and steadily more stable formulations of microbial biological control agents. It's interesting to note that growers quickly adopted the additional information and techniques needed to make biological control effective, and in many cases they developed new ideas and technology to enhance the release and establishment of invertebrate biological control agents. Additionally, they inspired scientists and the biological control sector to develop fresh invertebrate biological control techniques for newly emergent pests. When farmer organisations recognise the various benefits of biopesticides, including their economics, crop protection will undergo a new renaissance. They should take a far more proactive stance and demand expedited registration of cutting-edge sustainable control technologies in order to protect their own interests.

The market for biological control would significantly expand if the "real cost" theory were applied to chemical pesticides. Governments support the use of pesticides since the industry is not held accountable for human illnesses and deaths brought on by prolonged exposure to pesticides, nor is it required to pay for the cost of repairing environmental harm. As a result, costs associated with pesticides that have negative effects on human health and the environment are externalised and paid for by society, which is unethical and unscrupulous because the pesticide industry only benefits financially from these costs while bearing none of the responsibility. In the past, pesticides' profitability was in fact overstated. Chemical pesticide costs would increase significantly with realistic pricing that took into account true costs, and non chemical alternative controls would face fairer competition. Despite the fact that there have been known hidden costs associated with pesticides since the 1980s, prices of pesticides have rarely increased as a result. Applying levies on synthetic pesticides would be a first step toward true cost pricing because it would result in higher, more accurate costs for these products as well as more competitive pricing for the biological control agents employed in IPM programmes. (Pimentel and Burgess 2014; Bourguet and Guillemaud 2016; https://www.fortunebusinessinsights.com/thoughtleadership/biopesticides-trend-9099).

Supporting Points for Biopesticide Growth

1. Because of the rigorous battery of tests required for commercialization, some promising strains created by publicly supported research organisations in India are essentially confined to the shelf.
2. DNA bar-coding for precise identification of the species to be included in the creation of biopesticides before their field applications.
3. For the licensing and marketing of biopesticides in India, a comprehensive federal action plan, realistic budget, and efficient administrative procedures are required.
4. Farmers should receive sufficient training on using biopesticides in order to reap the greatest benefits.
5. The main drivers of the market expansion for biopesticides are the rise in demand for high-quality food, average entry hurdles, ecological imbalances, changing markets in developing nations, and ecological and health concerns for people, animals, and plants.
6. Environmental safety awareness raising, increased demand for chemical-free and environmentally friendly farming products, new product introductions, enhanced scientific validity of biopesticides, strict regulatory pressure, increased demand for organic products, and higher user confidence
7. Due to the efforts of Government of India programmes like *Paramparagat Krishi Vikas Yojana* (PKVY), Mission Organic Value Chain Development for North Eastern Region (MOVCDNER), and National Programme of Organic Production (NPOP), the area under organic cultivation increased from 1.5 million ha in 2016 to 1.9 million ha in 2018. A total of 1.35 million metric tonnes of organic food

were produced in the nation in 2016 by 0.65 million organic farmers. The biopesticide market in India is being driven by a sizable organic farming sector, and this trend is expected to continue (https://www.researchandmarkets.com/reports/5175605/india-biopesticides-market-growth-trends-and#rela2-5214644).

8. The 'Green Revolution' is progressively giving way to the 'Ever-green Revolution', especially in countries with strong agricultural foundations. Due to their eco-friendly, economical, farmer- and consumer-friendly qualities, the role of agri-bio inputs will thus be vital in fostering this shift. Additionally, it is anticipated that growing consumer demand for foods (organic food products) free of synthetic pesticides would further fuel market expansion.
9. Growers are becoming more and more eager to try biological solutions when conventional synthetic alternatives lose their effectiveness as a result of biotic stressors acquiring resistance.
10. The prohibitively high costs of developing synthetic crop protection chemistries are another factor driving the development of biopesticides. According to tech developers, the development and approval of a novel synthetic pesticide typically takes \$250 million and nine years, whereas a biopesticide requires less than \$10 million and four years.
11. The proliferation of start-ups in the biopesticide industry has produced a highly competitive and cutting-edge environment for advancements. Synthetic pesticides have not seen as much recent release as innovative biopesticide components.
12. The Central Insecticide Board and Registration Committee, GOI, provided straight forward and quick regulatory approval for the registration of biopesticides.
13. The government of India banned 18 active chemical compounds in response to growing environmental concerns and household awareness of food safety.
14. A lot of Indian export goods don't match the required minimum residue level
15. Biopesticide usage is scaled up owing to government support and increasing awareness about the use of non-toxic, environment-friendly pesticides.
16. A long-term collaboration agreement for the sale of biological products from Bioworks Inc., including biopesticides, in India and South Asia.

17. The market is primarily driven by the issue of chemical pesticide residues, the appeal of organic farming, environmental concerns, and the simple registration procedure.
18. New applications for biocontrol, such as nanotechnology, RNAi, etc. that are not achievable with synthetic crop protection are what are causing the expansion of biopesticides.
19. Concrete proof of biopesticides' effectiveness in reducing crop damage and the resulting rise in crop yield
20. Affordable, high-quality items are readily available.
21. Supply chain management needs to be improved in order to use biopesticides more frequently. An effective distribution mechanism for biopesticides from the plant where they are produced to the farm where they are used is crucial in this regard.

Factors Restrain Biopesticide Market

1. Due to federal and state initiatives, the demand for biopesticides has increased, which has resulted in "driving the marketing of fake biopesticides."
2. The limited production of biopesticides with biopesticidal formulations, registered under the 1968 Insecticide Act
3. The expense and lengthy licensing process for biopesticides in India discourage businesses from investing in the study and development of biological pesticides.
4. Before registering and propagating biopesticides, it is necessary to confirm the microorganism's bio-safety. In order to do rigorous safety and allergy tests, many universities and research institutes who conduct the original research and create biopesticides are unable to cover the additional costs. For instance, immune-compromised people have reported allergies to various fungi, such as *Trichoderma, Metarhizium, Anisopliae,* and *Beauveria*.
5. They cannot be employed against a variety of pests since they are target-specific, which is a limitation.
6. Biopesticides' effectiveness varies from climate to climate and is also extremely dose-dependent.
7. Due to the wide variations in the active and related substances of the parent plants in different agro-climatic zones, it is frequently challenging to make pure botanical pesticides, in contrast to synthetic pesticides, which can be made in desired purity and yield. Their physical and

chemical characteristics, as well as toxicological and other relevant features, change as a result. Their contamination by physical, chemical, or microorganisms also makes things more difficult.

8. The main market restrictions are farmers' lack of awareness and biopesticides' expensive price.
9. The market's expansion may be constrained by the lower acceptance rate of biopesticides than that of chemical agri-inputs.
10. Key heavyweights are strongly represented in the traditional and conventional agri-inputs sector, which is well-structured globally. But the biopesticides business is characterised by a number of start-ups that are having trouble getting enough money, building the right infrastructure, and getting traction with customers.
11. Many smaller developers may find it difficult to compete with established, potent synthetic pesticides, both in terms of proving the effectiveness of biopesticides and, more crucially, in persuading producers to switch from their tried-and-true ways to new and somewhat unproven products.
12. In addition, several types of biopesticides, particularly those produced from genetic materials or crop diseases, face unknown regulatory approval paths; this increases the difficulty of licensing and commercialization and impedes the innovation and development of biopesticides.
13. The bulk of biocontrol strategies necessitate repeated, frequent treatments for best results. These applications require more work and money, which sometimes acts as a barrier and stunts the development of biopesticides.
14. The research and development (R&D) of biopesticides carried out by small businesses consistently fails to understand the demand dynamics of a given location, which can further impede market growth.
15. The development of biopesticides is a high-risk business since it requires an initial large capital outlay to choose prospective strains for sales, as well as packaging, storage, and distribution.
16. Aside from the aforementioned problems, the single biggest barrier to the development and growth of biopesticides is the widespread selling of substandard (low CFU count), fake (no CFU count products) (Alam 1995). and misbranded (pesticide-laced bioproducts-pseudo-biopesticides) biopesticides (Keswani et al. 2016). APEDA (Ministry of Commerce)-certified organic bio-inputs supplied under the pretence of uncontrolled organic bio-inputs (not permitted by CIBRC) also constitute a severe threat to high-quality biopesticides. The organic bio-input products are not put through any bio-efficacy/safety experiments as required by

CIBRC. These categories account for over 65% of overall biopesticide sales (Singh and Arora 2016). To make biopesticides an effective tool for IPM/Sustainable Agriculture, the agriculture departments should strictly enforce the licensing requirements and quality controls for them.

17. The high prices for biopesticide registration (http://ppqs.gov.in/divisions/cib-rc/guidelines) are another barrier to advancing research, development, and usage of biopesticides in IPM/Sustainable Agriculture. Relaxed rules for CIBRC registration should be drafted in order to enable the registration of many biopesticides (GHS). The main barrier to promoting the use of biopesticides, biofertilizers, and botanicals is the imposition of a 12 percent Goods and Services Tax (GST) on microbial goods as well as on the botanical product (neem) (the same GST rate as for toxic/hazardous conventional chemical pesticides).
18. Farmers have a lot of concerns about the short shelf life of biopesticides. Because live bacteria make up the majority of biopesticides, changes in temperature, humidity, or even exposure to ultraviolet radiation reduce their effectiveness (Arora et al. 2016). Additionally, contamination may significantly lower the product's microbial count, greatly decreasing its efficacy in real-world settings (Alam 2000; Evans et al. 1993). Due to a shortage of money for the next steps, everything stalls before: adherence to regulatory requirements, scalability for application and delivery, marketing, and commercialization.
19. The rapid emergence of the corona virus pandemic has had an impact on the world market as countries have implemented lockdown measures and restricted public movement. These activities are having a substantial impact on the manufacturing of biopesticides, as firms are experiencing supply chain disruptions, a shortage of raw materials, forced plant closures, and a lack of staff.

Conclusions

Long-lasting shelf of biopesticide formulations, DNA-bar-coding for precise identification of organism, comprehensive federal action plan, training and awareness, rise in demand for high quality food, strict regulatory pressure, increase of area under organic farming, Green revolution to Ever-green revolution, proliferation of start-ups of biopesticide units, quick regulatory approval for registration by CIB&RC, New Delhi, India, withdrawal of dangerous chemical pesticides, negative consequences of Green revolution, failure of novel applications such as nanotechnology, RNAi etc., in development synthetic plant protection solutions etc., are considered as few driving forces to boost the growth of biopesticide in Indian agriculture. The USEPA has implemented a number of measures for the quick expansion of the biopesticide market, including appropriate changes to the registration process for the quick commercialization of biopesticide formulations, real cost policy, stakeholder perspectives, manufacturer and dealer of biopesticides, etc. Sustainable agriculture is referred to as "Conscious Agriculture" and is positioned between contemporary agriculture and conventional agriculture. All parties involved in the production and consumption processes must participate in conscious agriculture, which also protects the environment and the availability of resources for future generations. According to a position paper Fresco and Poppe recently issued, conscious agriculture is smoothly integrated into a "shared agricultural and food policy" (2016).

References

Alam G (1995) Biotechnology and sustainable agriculture: Lessons from India. OECD development centre, working paper no. 103, Sustainable Development: Environment, Resource Use, Technology and Trade. doi:10.1787/711068780307

Arora NK, Verma M, Prakash J, Mishra J (2016) Regulation of biopesticides: Global concern and Policies, 283-299pp. In: NK Arora et al. (eds), Bioformulations for sustainable agriculture. Springer India

Bourguet D, Guillemaud T (2016) The hidden and external costs of pesticide use. In: Lightfouse E (ed) Sustanable agriculture reviews. Springer, Dordrecht, 35-120pp.

Calvo FJ, Bolckmans K, Belda JE (2021) Biological control-based IPM in sweet pepper greenhouses using *Amblyseius swirskii* (Acari: Phytoseiidae). Biocontrol Sci Technol 22:1398-1416.

EC (2009) Sustainable use directive. European Parliament and of the Council of 21 October 2009establishing a frame-work for community action to achieve the sustainable use of pesticides. Off J Eur Union L 309:71-86.

Evans J, Wallace C, Dobrowolski N (1993) Interaction of soil type and temperature on the survival of *Rhizobium leguminosarum* bv Viciae. Soil Biol Biochem 25:1153-1160

Fresco L, Poppe K (2016) Towards a common agricultural and food policy. Wageningen University and Research, Wageningen. doi:10.18174/390280.

Keswani C, Sarma B, Singh H (2016) Synthesis of Policy Support, Quality Control, and Regulatory Management of Biopesticides in Sustainable Agriculture. 10.1007/978-981-10-2576-1_1.

Pimentel D, Burgess M (2014) Environmental and economic costs of the application of pesticides primarily in the United States. In: Pimentel D, Peshin R (eds) Integrated pest management. Springer, Dordrecht, 47-71pp.

Singh R, Arora NK (2016) Bacterial formulations and delivery systems against pests in sustainable agro-food production systems against pests in sustainable agro-food production. Module in Food Sci, Elsevier, 1-11pp.

Urbaneja A, Gonzalez-Cabrera J, Arno J, Gabarra R (2012) Prospects for the Mediterranean basin. Pest Man Sci 68:1215-1222.

van Lenteren JC, Bolckmans K, Kohl J, Ravensberg WJ, Urbaneja A (2018) Biological control using invertebrates and microorganisms: plenty of new opportunities. BioControl 63:39-59.

Links

https://www.fortunebusinessinsights.com/thoughtleadership/biopesticides-trend-9099

https://www.researchandmarkets.com/reports/5175605/india-biopesticides-market-growth-trends-and#rela2-5214644